BK 629.892 C678R
ROBOT TECHNOLOGY
C1983- 42.95 FV
3000 684804 30016
St. Louis Community College
W9-COT-830
WITHDRAWN

St. Louis Community
College

Robot Technology

Series Editors and Consultants: Philippe Coiffet and Pierre-Jean Richard
English Language Series Consultant:
Professor I. Aleksander, Brunel University, Uxbridge, England

Volume 1

MODELLING AND CONTROL

Philippe Coiffet
Automation Laboratory, Montpellier, France

Kogan Page
London

Prentice-Hall, Inc.
Englewood Cliffs, NJ 07632

Library of Congress Cataloging in Publication Data

Coiffet, Philippe. (date)
Robot technology.

Translation of: Les Robots.
Includes bibliographies and index.
Contents: v. 1. Modelling and control.
1. Automata. 2. Robots, Industrial. I. Title.
TJ211.C6213 1983 629.8'92 83-3091
ISBN 0-13-782094-1 (v. 1)

Production Supervision: Karen Skrable
Manufacturing Buyer: Gordon Osbourne
Cover Design: Wanda Lubelska
Cover Photo: Courtesy of Cincinnati Milacron

Printed in the United States of America

10 9 8 7 6 5 4 3

ISBN 0-13-782094-1

Prentice-Hall International, Inc., *London*
Prentice-Hall of Australia Pty. Limited, *Sydney*
Editora Prentice-Hall do Brasil, Ltda., *Rio de Janeiro*
Prentice-Hall Canada Inc., *Toronto*
Prentice-Hall of India Private Limited, *New Delhi*
Prentice-Hall of Japan, Inc., *Tokyo*
Prentice-Hall of Southeast Asia Pte. Ltd., *Singapore*
Whitehall Books Limited, *Wellington, New Zealand*

Contents

Series Foreword

This first volume in the series is based on part of a course that I taught in various state and private institutions during 1979 and 1980. It was concerned with the design and control of stationary, articulated robots operating as non-feedback systems. This book (which is devoted to the statement and understanding of problems rather than their solution) presents a review of some of the work, started in 1972, of the team involved with robotics and biomechanics at the Automation Laboratory at Montpellier, in association with the French National Centre for Scientific Research. I am greatly indebted to Professor A. Liegeois, the guiding light of the team, to A. Fournier, E. Dombre, W. Khalil, P. Molinier and P. Borrel who have contributed so much to the progress made in this field.

The second volume will describe the present state of the knowledge of robotic systems which are able to perceive the environment, to a greater or lesser extent, and to react appropriately. Such robots will be able to perform the function required of them, in spite of unpredictable, limited changes in the environment.

The third volume will deal with teleoperations, the branch of robotics characterized by the necessary and almost permanent presence of a human operator at the controls. The fourth volume will be concerned with technological components of robots and further volumes will discuss robotic languages and programming methods, decision autonomy and artificial intelligence and, finally, the computer-aided design of robots.

Philippe Coiffet

Preface

Robotics is much discussed, both in the popular press and in the technical journals. There is widespread feeling that it is likely that robots, in the years ahead, will become crucial agents of industrial change, transforming production processes and changing the lives of millions. What makes the industrial robot such a powerful tool? The main difference between it and the conventional machine is that the robot is able to react with its environment, ie it is an *adaptive* machine. This is clearly different from other machines, which can be used only for well-defined, specialized and pre-appointed tasks. The robot is able to react to changes in its environment because, like the computer, it has an in-built, software-controlled intelligence.

The publicity surrounding the introduction of robots into industrial situations, however, exaggerates the true state of the theoretical and practical knowledge of robotics. The overall concepts are more easily understood than the theoretics and practicalities involved in producing actual components.

No one doubts the feasibility of building a manipulator arm having a performance comparable with the human arm. Yet the technical challenge is considerable because, at present, the artificial 'arm-eye' device is far from attaining the standards of performance required in industrial operations. The majority of the 50 000 robots in use today are essentially simple machines which reproduce a movement, and are unable to react with even predictable changes in the environment — they are without the capacity to detect the complex information directed to them from the environment.

These difficulties explain the slow progress made in robotics research. Laboratories with an interest in automatic control systems have not given a great deal of attention to robotics. However, a few have attempted to introduce robotics into their field of study, in particular in the areas of aerospace and biomedical research. Compared with other countries, France had, before 1975, an almost unique number of workers involved in the study and development of robotics. At the forefront was the Automation Laboratory at the University of Montpellier, where the author of this book is an active research worker.

The team at Montpellier is involved in the study of the normalization and control of articulated systems. This is one of the most difficult and intractable of subjects since, to date, most fundamental research has been done into subjects such as artificial intelligence and pattern recognition. Yet robots are intended to perform specific tasks in actual working environments and, without manipulator competence, they can have no industrial role. Articulated systems are difficult to study because they possess multiple degreees of freedom and also because they are essentially non-linear in character. For the research worker interested in automatic control these systems are just as important as the physicochemical processes and aeronautical systems which have been the subject of so many detailed studies.

It is with great pleasure that I welcome this volume in the series as one of the first fundamental monographs on the subject. The depth at which the subject is treated will ensure its place as a source from which all future works will be able to draw. None of the essential foundations of the theory of articulated systems has been neglected, and the book provides a basis for the generalized treatment of methods of rational study for any robotic structure.

Readers unfamiliar with the mathematics of systems possessing several degrees of freedom should not be discouraged by the complexity of the equations described in this series. They can easily be solved with the use of a computer, following the methods developed by the author.

This book, concerned as it is with concepts of robot modelling, provides for the first time a treatment of their dynamic control. In the future, robots will be available which will be able to act quickly and precisely and which will adapt automatically to changes in the environment.

With this, the first volume of *Robot Technology*, we have available for the first time the basic knowledge needed by designers and operators of robots, the knowledge they will need for a proper understanding of the fascinating developments which are sure to arise in this new and exciting industry.

Jean-François Le Maître
Director of Automation, Plant & Technology,
Renault Industries

Chapter 1
Definitions and objectives

1.1 Origin of the word 'robot'

The word seems to have first become popular when Karel Capek's play, *Rossum's Universal Robot*, was performed in France in the 1920s. In the play, small artificial and anthropomorphic creatures strictly obeyed their master's orders. In Czech these creatures were called 'robota' (which is the Russian word for 'work').*

1.2 Robotics today

Robotics can be defined as the theory and practice of automation of tasks which, because of their nature, were previously thought to be reserved for man alone. Such work is characterized by an almost permanent interaction between the robotic device and the object (or environment). Implicit in such interaction is some kind of pre-appointment of the task.

It would appear that adaptive execution of this kind would call for the use of an operator's reflexes and intelligence; this is why the word 'robotics' is often linked with the notion of artificial intelligence.

Bearing in mind the definition of robotics, three main categories can be identified in the current research being conducted into robotics. These are:

1. Research work on *individual robots*, situated in a fixed location or on a carrier vehicle (mobile robot).
2. Research work on robots operating in conjunction with other robots or other machines. Such production lines constitute what can be termed *flexible manufacturing processes*.
3. Research work on *teleoperation* (work controlled from a distance). In such processes a human operator must be present at the controls of the machine because the job can neither be pre-programmed

*The reader with an interest in the folkloric history of robotics is referred to *Human Robots in Myth and Science* by J. Cohen, George Allen & Unwin, 1966.

nor executed automatically in the adaptive mode. This is because machines are not yet available which can analyse and interpret their environment when their application is changed from one function to another. In teleoperation systems some degree of human operation is still required and the robot acts as an aid to the operator rather than a replacement for him.

There are three main types of such systems:

1. The acquisition and presentation of relevant and easily interpretable information to the operator. Examples of this include the presentation of the stereoscopic view of a gripper, or the indication of the forces existing between two components that are to be fitted together in an assembly process.
2. The automatic monitoring of an operator's movements and the provision of starting signals which activate the interruption of transmission from master to slave when the precision of the operator is failing. This function is also concerned with a system's self-testing facilities.
3. The automation of various functions – so freeing the operator. These might include the gripping of objects when a device is in an automatic mode, or the maintenance of the horizontal when grippers are used to handle fluids, irrespective of the motion associated with their handling.

A specialized branch of robotics, with problems which are peculiar to it, is *medical robotics*. In the design and manufacture of devices such as prostheses and other manipulators, the same principles are used as those used in the construction of robots. However, adaptation of the principles of robotics to the problems of the handicapped, which are specific and often difficult to overcome, has achieved only limited success.

1.3 What is a robot?

1.3.1 TOOLS AND ROBOTS

As our ancestors became conscious of their environment they recognized the need to master it and began to invent tools. By making tools, primitive man increased his effective physical and territorial capacity. With the discovery of the lever the effect of the physical force that man was able to apply directly was increased. Similarly, the hammer helped him to perform tasks which previously would have been impossible; the pincer gripper enabled him to manipulate objects from a distance (particularly useful when using fire, for example).

Much later, when man invented machine tools that possessed a degree of *autonomy*, his role changed to that of inactive custodian – a role that, in essence, led him merely to start, stop and watch the machine at work. With the introduction of automation, this has been reduced to starting the machine by pressing a button which, at the same time, switches on sophisticated systems of control.

In addition to having a degree of autonomy, machines are now required which can produce action from a distance and are flexible in their application. The robot satisfies these criteria – it is an adaptive, powerful and autonomous tool.

1.3.2 SPECIFIC PROPERTIES OF A ROBOT

From the statements made in preceding sections two very important characteristic features emerge:

□ *Versatility:* The versatility of a robot depends on its geometrical and mechanical capacity. This implies a physical capacity to perform varied functions and to produce a diversified execution of simple tasks. Versatility also implies that every robot must have a structure with a geometry that can be modified if necessary.

□ *Auto-adaptivity to the environment:* This concerns a robot's potential for initiative in carrying out tasks which have not been completely specified and despite unforeseen changes in the environment. In this capacity the robot uses:
- its ability to perceive the environment (by the use of sensors).
- its capacity to analyse the task-space and to execute a plan of operation.
- its modes of automatic command.

Versatility is found to varying extents in most robots. It should be noted here that versatility does not depend on the number of degrees of freedom (DOF) alone, although in general an increase in the number of DOF does increase the level of versatility. There are other factors to be taken into account; in particular, the structure and capabilities of the end effectors (for example, can they be adapted to take different tools?).

The geometric structure of a robot may be determined by an analysis of the tasks it is to perform; however, it must be stressed that it will never be possible to define the entire range of tasks that a particular robot could perform. This point will be examined in detail in Chapter 10. Several problems remain:

1. The lack of adequate sensors for the acquisition and pre-processing of information received from the environment, particularly visual information.
2. The state of development of scene analysis and artificial intelligence

algorithms is not yet fully developed.
3. The slowness of the computations involved.

1.3.3 GENERAL STRUCTURE OF A ROBOT AT WORK

1.3.3.1 Constitution of a robotic system

A robot can only be sensibly defined in terms of the environment that it modifies and it is possible to identify four interactive parts (see Figure 1):

1. The machine equipped with actuators: Such machines typify the image we have of a robot. It is this type of machine which is designed to perform a specific task. For instance, the machine might be an articulated mechanical structure possessing several DOF. Commonly, six DOF are present in a robot designed to move objects. The first three DOF direct the gripper to a required position and the remaining three are used to orientate the end effector (see Figure 2).

The different articulations are controlled by actuators, very often electrical or pneumatic when dealing with small loads (a few decanewtons), and hydraulic for heavy loads (up to several hundred decanewtons).

2. The environment: These are the surroundings in which the machine is placed. For robots in a fixed position, the environment is reduced to the space which is actively researched by the robot, ie the volume swept out by the end effector when directed through all

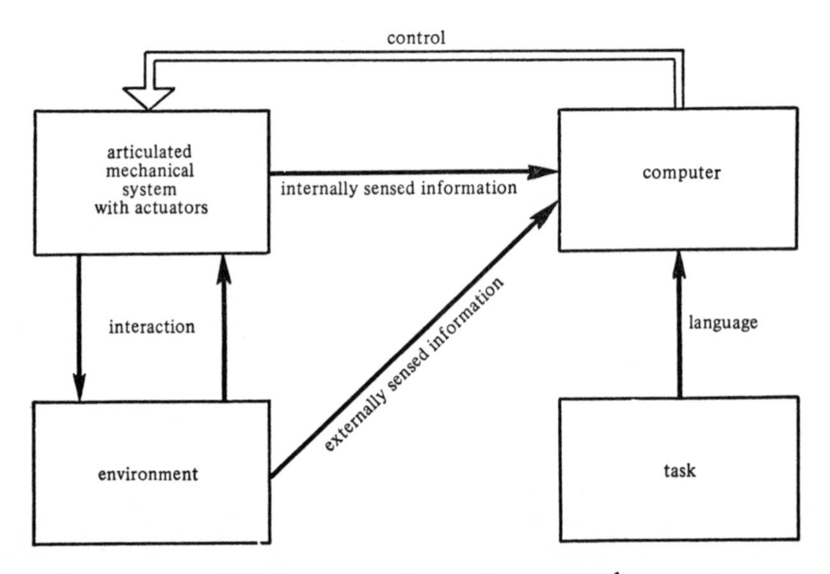

Figure 1. *Elements of a robot at work*[1]

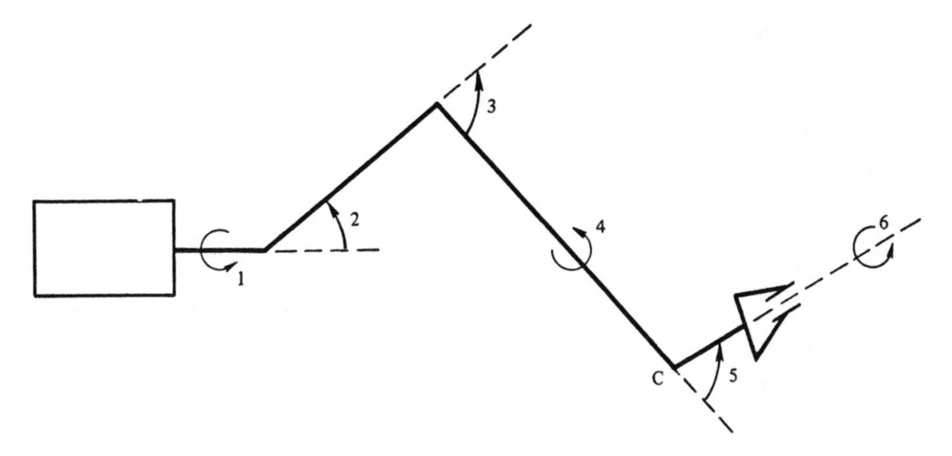

Figure 2. *Geometrical structure of a robot. The first three DOF are used to position point C (the gripper) in space; the last three DOF are used to orientate the gripper in any direction*

possible positions. It should be noted that the environment is not only defined by the geometrical considerations (ie 'reachable' space), but also by all the physical properties of the surroundings and everything that the surroundings contain. The nature and behaviour of a robot depends on both of these properties and also on the interaction between it and the environment.

3. The task: A task can be defined as the difference between two states of the environment – the initial state and the final state after completion of the task. The task must be described to the computer of the robot in an appropriate language. The description can have several forms and may even change during the course of an operation.

4. The computer or 'brain' of the robot: This is the part of the robot that generates the control signals (these signals actuate articulations of the robot limbs) according to *a priori* information (prior knowledge of the task to be performed), and *a posteriori* knowledge of both the present and past states of the robot and its environment. In simple cases the computer does not have to be digital but from now on it will be assumed that the robots considered here are under computer control.

1.3.3.2 How the system operates

Essentially the computer is provided with information representing a model of the robot, with details of the environment, data relating to the tasks to be performed and with a number of planning algorithms. When in operation it receives at all times information concerning the robot (*internally sensed information*) and the environment (*externally sensed information*). By using this information in conjunction with planning algorithms, which can refer back to past experience, the computer develops its control over the robot, causing it to move towards the correct execution of the task assigned to it.

1.3.4 NATURE OF A CURRENT INDUSTRIAL ROBOT

A typical industrial robot does not have the capacity to initiate action. All the necessary sequences of movement are determined beforehand and then all the required information is fed into the computer of the robot in the form of pre-recorded programs. If the effect of the sensors (end- or proximity-switches) which provide logical on-off information (and which are used for safety purposes) is excluded, it is apparent that such a robot cannot adapt to the environment.

To represent a contemporary robot Figure 1 has been simplified to that shown in Figure 3. In this, an environment-robot relationship is not evident. This illustrates the importance of a *static* or *perfectly reproducible* environment described inside the computer.

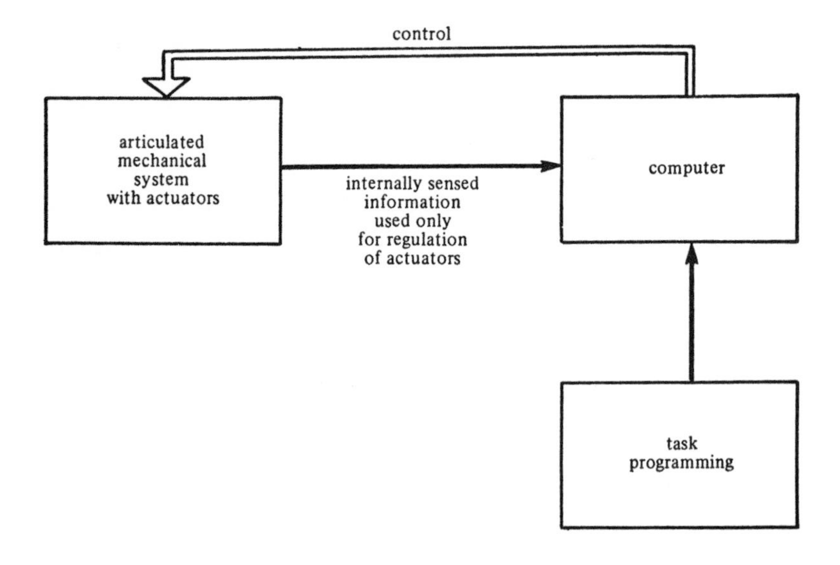

Figure 3. *The functioning of an operational robot*

1.4 Classification of robots

There are two classes of robots:

1. Pre-programmed robots which cannot examine the nature of the

task they are performing. These are controlled by programs which are not self-modifying in their mode of operation. Most industrial robots are of this kind (non-heuristic).
2. Robots capable of adapting at least part of their behaviour in response to information of their environment received by their sensors. These are currently at an experimental stage.

There are two main types of robot classification in use at present. The first is concerned with physical characteristics,[2] ie robots with four DOF, robots having a load capacity in excess of 100 kg, robots controlled by a programmed system, or those having a certain architectural form. This classification has practical advantages, because the needs of the user are clearly expressed. It is possible to organize a system of classification with taxonomic principles similar to those used in the biological sciences[3] and geology. A decision tree would then allow a user to find a robot appropriate to his needs.

Another method of classification would be to attempt to classify robots according to generation,[4] indicating say generation 1, 2, 3 or 1, 5 or 3, 5. This system could be criticized because of the lack of knowledge of the relationship existing between the capacity, make-up and intelligence of a robot.

Conclusions

In this first chapter, what is meant by the terms 'robot' and 'robotics' has been explained in a general and non-mathematical way. This volume will deal with a common industrial robot (Figure 3) for which interaction with the environment can be pre-programmed but not changed during operation.

The following words will be used arbitrarily as synonyms: *robot manipulator, mechanical system, articulated mechanical system, articulated robot.*

Chapter 2

Structure and specification of articulated robots

The geometrical structure of a robot varies with job specification. The factors that dictate the physical make-up of an articulated robot, both from a theoretical and a practical point of view, are considered in this chapter.

2.1 Degrees of freedom of a solid

Consider a solid placed in real three-dimensional space. This solid possesses six DOF, three for movement and three for orientation. If an orthonormal trihedral is considered, which is centred on the centre of gravity of the solid, the six DOF can be expressed by the three axes of the trihedral so they perform a movement. The other three DOF can be 'used' to perform a change in orientation, ie when producing rotations with reference to the three axes of the trihedral.

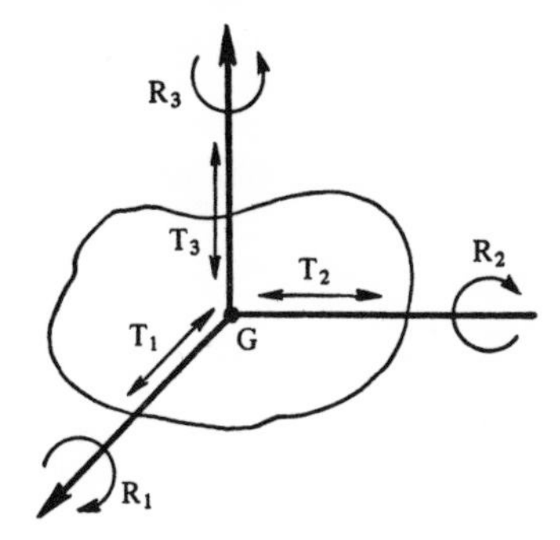

Figure 4. *The six DOF of a solid: three translational (T_1, T_2 and T_3) and three rotational (R_1, R_2 and R_3)*

2.2 Degrees of freedom of a robot

The function of a robot is, as has already been seen, to bring about a specific local modification of its environment, using tools it possesses or is capable of using, when commanded. Three substructures can be distinguished in the overall geometric structure. These are interconnected and often mechanically coupled.

1. The vehicle: The robot must reach its operating location. Take the example of a manipulator used in the maintenance and repair of spacecraft. The manipulator itself is fixed to the spacecraft which will move to (three translational DOF) and then orientate itself to the object to be repaired so that the robot may operate (three rotational DOF). In almost all cases the vehicle will possess six DOF. The vehicle generally provides one translational DOF. In many industrial situations the robot is not orientated at all (eg when used to paint the hull of a ship or to move radioactive materials) and the task of orientation is taken on by the operator whilst the vehicle often takes the form of a moving gantry, capable of producing the three translations necessary for movement (without the need for the three DOF).

2. *The arm:* The arm manoeuvres the end effector to a precise location. It must possess three DOF and its action can thus be realized by using rotational as well as translational actions. It is known that a rotation about the axis of a solid, not passing through its centre of gravity, has a secondary effect, ie the translation of the solid.

These three DOF can be called the three primaries in the sense that they start at the base (the part joined to the vehicle), finish at the end effector, and are encountered in succession.

3. *The end effector:* The arm, having positioned the end effector in the correct location, has only to be correctly orientated. To obtain any possible orientation of the end effector, three rotations about three normal (or possibly converging) axes need be made.

For practical reasons, (concerning the limits of a robot's reach) to perform any imaginable movement, the articulated system should possess at least nine and perhaps 12 DOF (Figure 5).

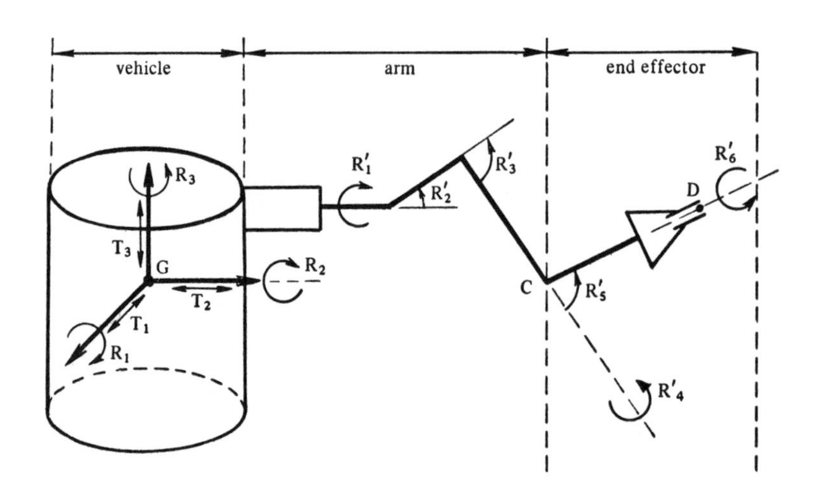

Figure 5. *An example of a robot, of the type used on a spacecraft, possessing up to 12 DOF*

2.3 Position of the vehicle and resulting redundancy of degrees of freedom

With reference to Figure 5, if it is supposed that a slight displacement at point C is required, using rotations R'_1, R'_2 and R'_3, then it is possible to make this movement using only the limbs illustrated. However, there are nine DOF available to act on C which implies that six are redundant. So, why not alter the positions of R_1, R_2 and R_3 or T_1, T_2 and T_3, rather than those of R'_1, R'_2 and R'_3? This possibility will be discussed again later. In practice the choice is of great importance, and depends on two considerations:

1. Precision of movement: If, to perform the movement, rotational DOF are used, it is obvious that for such a movement the longer the lever arm, the smaller the rotation necessary. The smallest controllable rotation has an easily attainable lower limit. If the translational DOF is used, the problem is one of lack of precision because of the distance involved, and possible vibration of too long a lever arm. This is why it is preferable to use actions that are directly above the point to be moved.

2. Function and control decoupling: Considering Figure 5, it is possible to propose a model having 12 variables (or 24 if a state vector is used). The equations are complex and involve high computational overheads. It would appear to be better to aim for decoupling between the DOF of the vehicle, the arm and the end effector, and to work on the individual substructures in the following way.

Once the vehicle is correctly positioned it should not be moved. In the same way, if the task can be related to the position of point C and the orientation of CD, it would be possible to operate using systems R'_1, R'_2 and R'_3 and R'_4, R'_5 and R'_6 (assuming them to be decoupled).

From here it is assumed that the vehicle is properly sited, allowing discussion of problems related to the function of a robot in a fixed position.

2.4 How many degrees of freedom?

For an articulated mechanical structure six DOF are adequate for most practical purposes. However, many such systems, eg those used for handling objects, have less than six DOF. The number of DOF is dependent on the character of the task. In Figure 6a, a manipulator with four DOF is illustrated, which can only be used to grasp objects in a vertical plane. A large number of industrial tasks do not require as many as six DOF for their execution. Equally, it is rare for an arm to possess less than three DOF. It is almost always in the end effector that some reduction in mobility is observed.

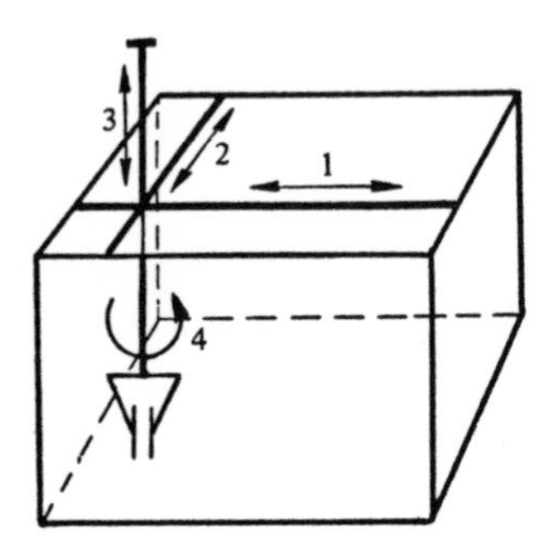

Figure 6a. *An example of a manipulator with four DOF*

Occasionally, more than six DOF are needed to overcome the problems caused by the need for precision positioning or by the special requirements of a job, eg the assembly of two parts of a cylindrical plug, using an articulated arm fixed on a vehicle. Taking as an example the assembly of an underwater fuel pipeline (see Figure 6b), the orders

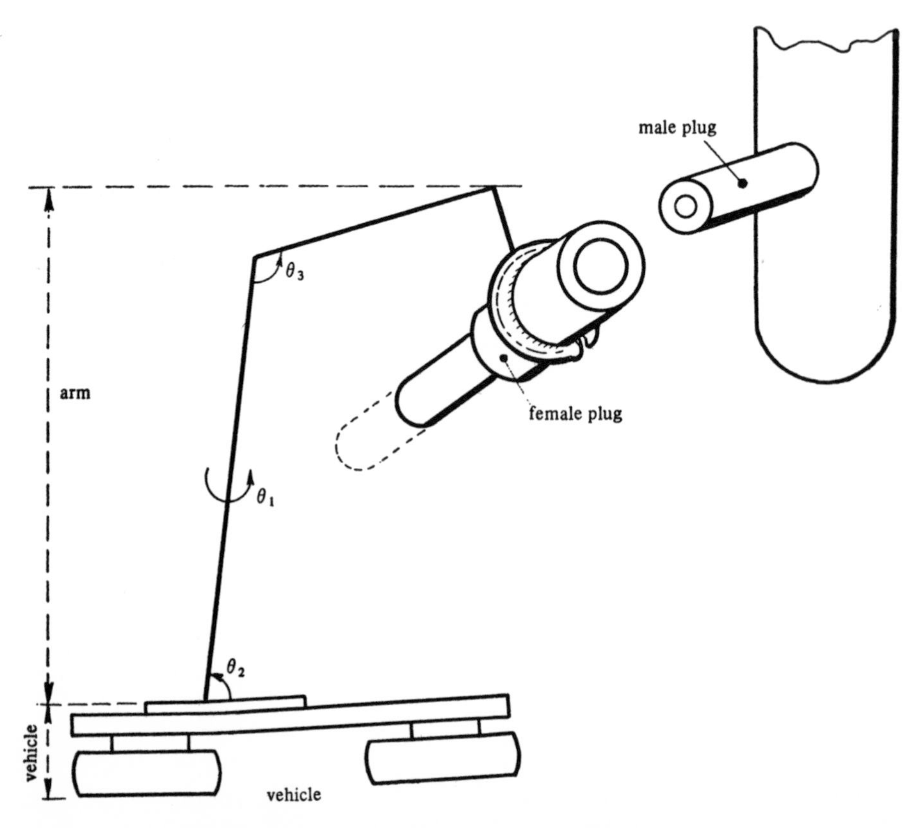

Figure 6b. *An example of a job requiring more than six DOF*

of magnitude of the loads involved are:

- ☐ mass of the part to be moved = 1 tonne
- ☐ outside diameter = 1 metre
- ☐ inside diameter = 0.5 metre
- ☐ difference between the diameters of the male and female parts to be assembled = 0.001 metre
- ☐ maximum stretch of arm = 6 metres

If the vehicle is brought to a suitable fixed position and orientation, with the arm and end effector having three DOF each, the six DOF obtained should be sufficient for the assembly of the parts (the assembly requires only five DOF if the symmetry of rotation of the component is taken into account). Thus, it would be possible to operate through angles θ_1 or θ_2 to produce movements (of a few tenths of a millimetre) of the female part during the assembly process. However, θ_1 and θ_2 would need to be controlled to an accuracy of a few seconds of an arc; this is hardly realizable as it is extremely expensive to achieve in practice.

A much better technique is to use both the arm and the vehicle to achieve assembly of the two parts. The end effector would then have five DOF, of small range, which could be controlled and the arm could be left in a fixed position to perform the connection. Without taking the vehicle into account, eight DOF would be needed.

To use a robot possessing several arms, more than six DOF are needed. For teleoperation applications two-arm systems are used[5] because two hands are needed; one grips and holds, the other carries out the task on the fixed components – such robots (or here 'bi-robots') possess 12 DOF.

2.5 False degrees of freedom

If the apparent mobility of a segment (or body) or part of a mechanical articulated system does not increase the number of DOF already present, this mobility is termed a *false DOF*. False DOF can be classified into three groups (two useful, the other of doubtful use).

1. The DOF of a tool: The end effector of a robot either constitutes a tool or is shaped to carry one. In order to function the tool must be free to move; a drill or a countersink must be free to rotate, a gripper must be able to open and close, and so on. Several manipulators have grippers as end effectors. It is clear that the extra DOF which results from the control of the opening or closing operation should not be classified in the same way as the DOF which ensures the movement and orientation of the gripper. Indeed, some commercial robots are wrongly classified in this respect by the manufacturer.

2. *The increase in the number of articulations needed to avoid obstacles:* In some applications, the articulated mechanical system must be able to access a narrow scuttle, or be able to operate behind a panel. The classical case of six DOF linked by the lowest number of rigid segments becomes inoperative. A flexible structure must then be used. The problem may be solved by increasing the number of segments and by making them mobile with respect to each other. However, the number of DOF is not necessarily increased. Figure 7 shows the skeleton of a 'painter robot' which appears to have ten DOF but in reality has only six, because the following relationships still apply:

$$\theta_4 = \theta_6 = \theta_8 \text{ and } \theta_5 = \theta_7 = \theta_9$$

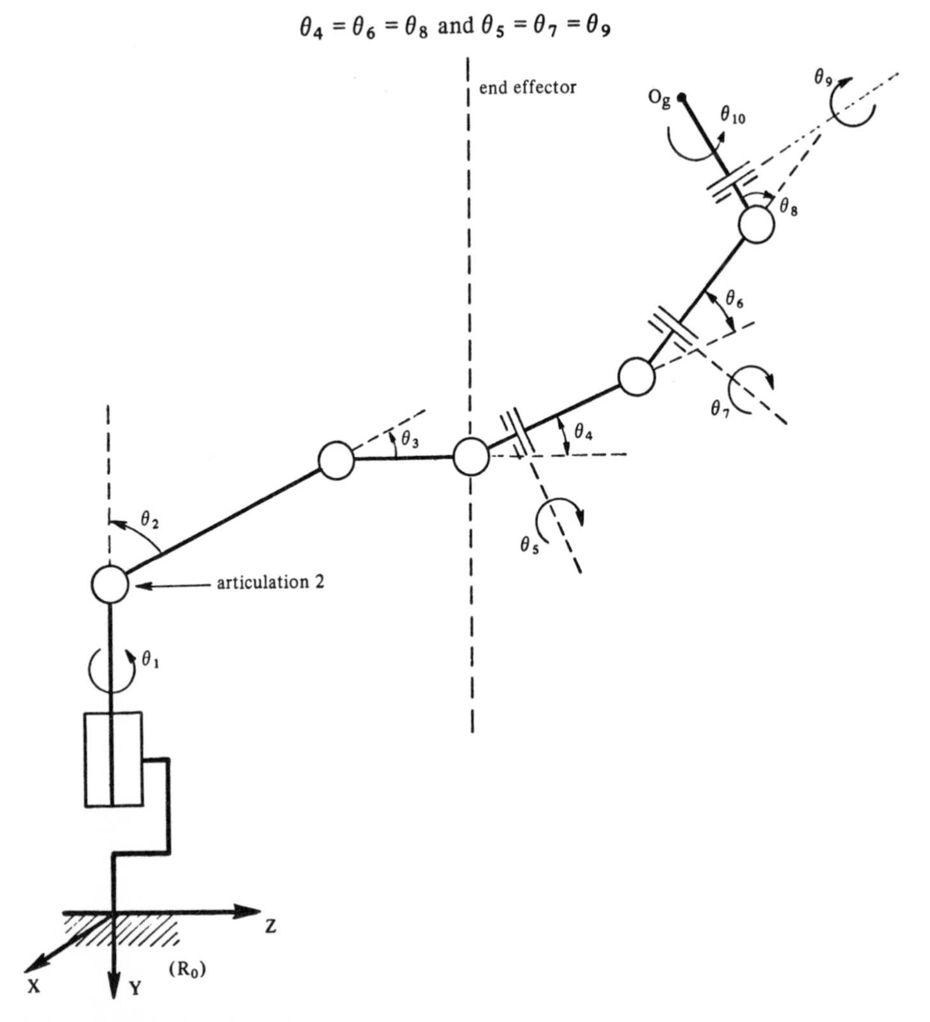

Figure 7. *A robot having ten articulations but possessing only six DOF:* $\theta_4 = \theta_6 = \theta_8$ *and* $\theta_5 = \theta_7 = \theta_9$*; such robots are typically used in paint-spraying applications*

3. Compensation for the lack of DOF on the end effector by the creation of upper DOF on the arm of the vehicle: It has already been noted that robots with less than six DOF are deficient at the end effector. The practical necessity of using the DOF closest to the robot's extremity (or load to be moved) has also been noted. It is clear that the creation of DOF must follow certain rules and that an increase is not always beneficial. Examination of the manipulator in Figure 8 shows it to have five DOF: three related to the arm and two related to the end effector (α_6). An extra DOF can be created by mounting the robot on a rail (α_7).

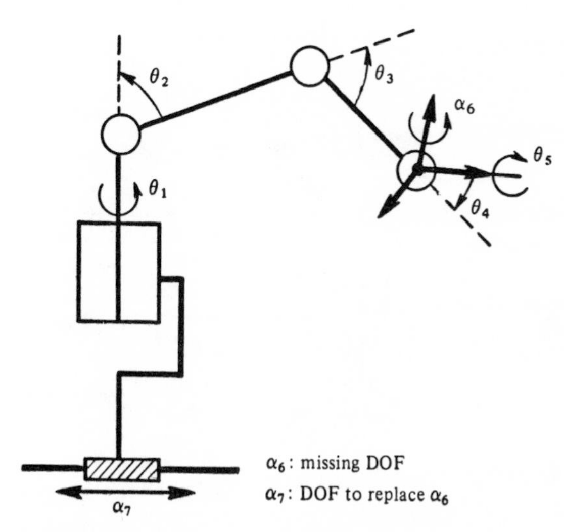

Figure 8. *The architecture of a robot possessing five DOF*

In these circumstances it should be noted that: α_7 is a DOF which belongs functionally to the vehicle, since θ_1, θ_2 and θ_3 belong to the arm. Replacement of α_6 by α_7 creates the problems discussed in section 2.3.

If the manipulator belonging to α_7, θ_1 and θ_2 of the arm is analysed, the end effector is seen to have θ_3, θ_4 and θ_5 as DOF – but θ_3 and θ_4 are rotations about parallel axes. The end effector does not always have a DOF equivalent to α_6. It is necessary to be aware of such false DOF as they are not always obvious and they can cause problems during use.

2.6 Architecture of the arm

A study of a robot's first three DOF indicates that the various combinations of rotation and translation can produce 42 different structures. It appears that, in practice, as Dombre[6] has shown in a study of 115 robots, only four or five combinations are used; Figure 9 illustrates these five combinations and indicates their percentage use.

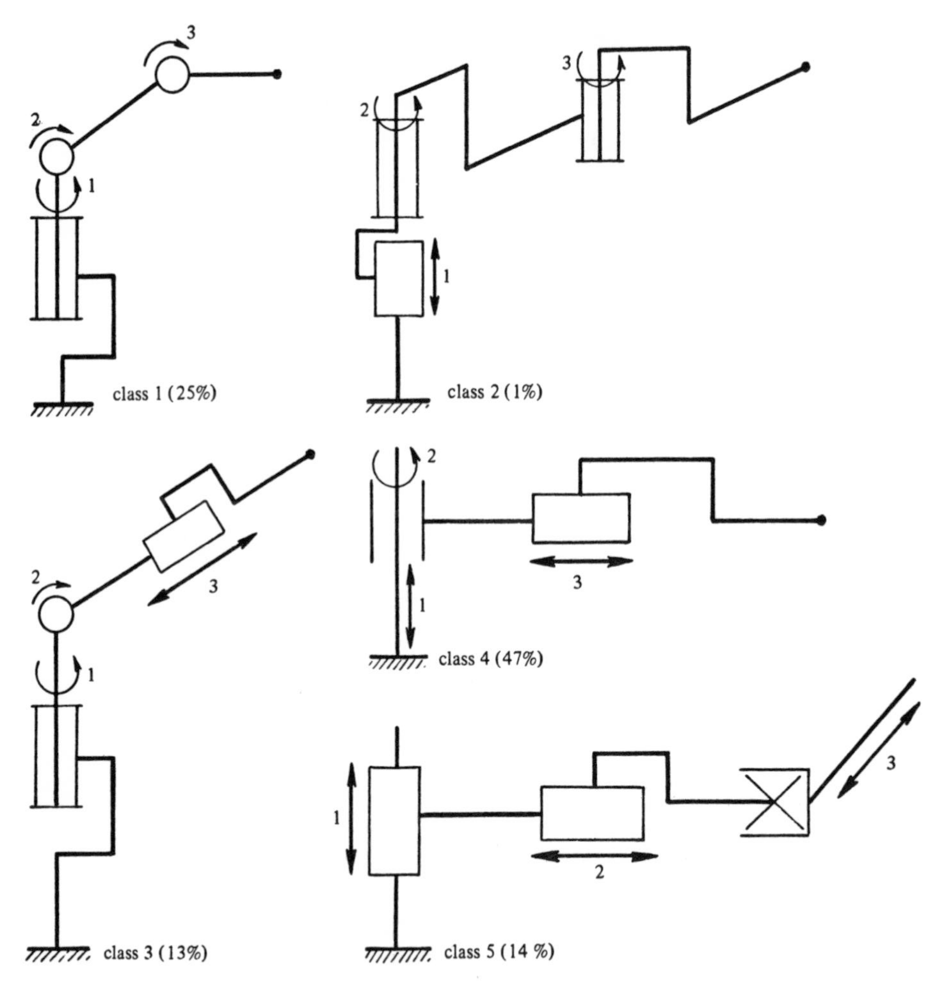

Figure 9. *The five main types of arm, with an indication of their percentage use*

2.7 Description of articulated mechanical systems

To achieve computerized automatic control in an articulated mechanical system (AMS), it is necessary to start with a model of such a system (see Figure 10).

This representation is described in terms of characteristic variables which are specific to the system and include DOF, lengths, masses, inertias, etc. The number and nature of these parameters vary and

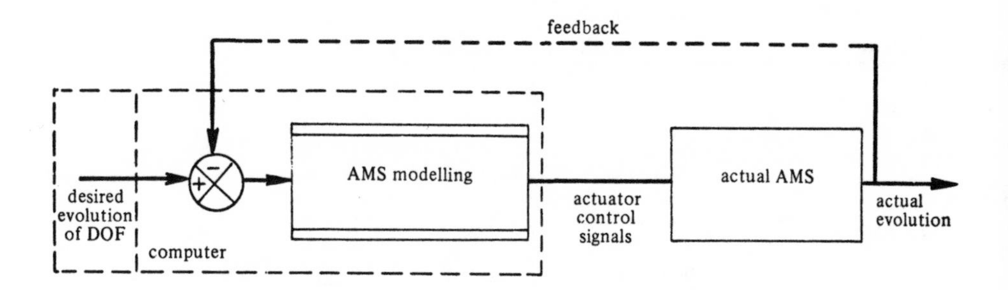

Figure 10. *Control of an AMS using a computer*

depend on whether the model is dynamic, kinematic or geometric. Together these variables constitute the AMS. The aim of this description is two-fold:

1. To be able to use the description for the software models; the most important consideration being the language used to describe the system to the computer.
2. To be able to use this description to compare robots, to recognize despite differences in their appearance, when they belong to the same class, eg one might be fixed to a wall, another fixed to the floor. Robots can look different but have the same combination of DOF. The numerous representations available have been described previously by Borrel[7] and from now on this classification will be the one used.

2.7.1 GRAPHICAL REPRESENTATION

Graphical representation of articulated systems is made difficult by the problems associated with the illustration of the different types of articulations – the standard representations are shown in Figure 11. In previous figures similar standards have been used, and arrows have been used for clarification. Throughout the book this system will be used freely, with the occasional addition of explanation. Such graphical representation could constitute the basis of a language suitable for the description of mechanisms and directly transferable on to a computer keyboard. However, to date, graphic languages of this kind have not been used in this way.

2.7.2 THE ROTH-PIEPER DESCRIPTION[8]

If it is assumed that the joint between two segments has only one DOF, either rotational (R) or translational (P), then starting at the end and progressing in the same order as the robot's DOF, for the five groups of

Type of articulation	Relative movements	Number of DOF	Symbols
Fixed beam	0 rotation 0 translation	0	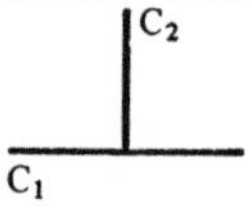 C_2 C_1 C_1 : body 1 C_2 : body 2
Pin	1 rotation 0 translation	1	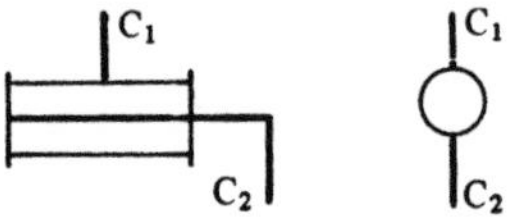 C_1 C_2 — C_1 C_2
Slide	0 rotation 1 translation	1	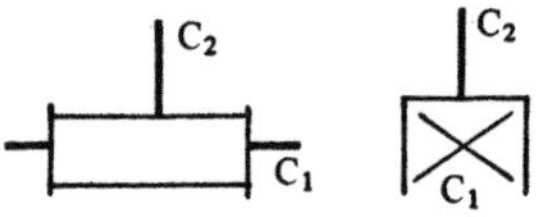 C_2 C_1 — C_2 C_1
Helical slide	1 rotation 1 twin translation	1	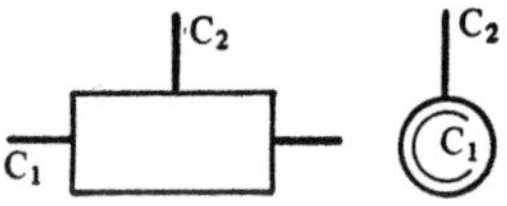 C_2 C_1 — C_2 C_1
Sliding pin	1 rotation 1 translation	2	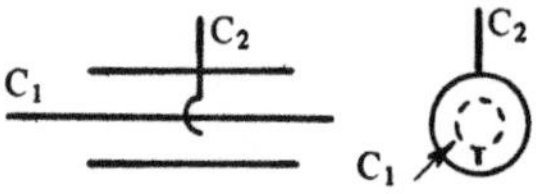 C_2 C_1 — C_2 C_1
Sill	1 rotation 2 translations	3	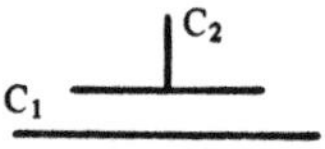 C_2 C_1
Swivel	3 rotations 0 translation	3	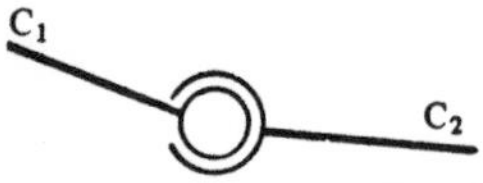 C_1 C_2
Linear joint	2 rotations 2 translations	4	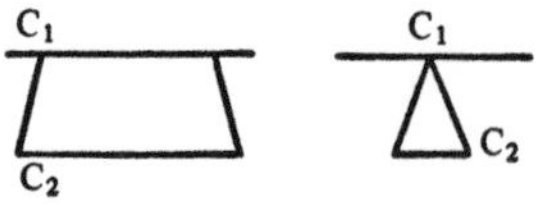 C_1 C_2 — C_1 C_2
Round joint	3 rotations 1 translation	4	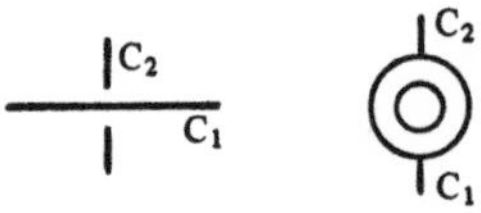 C_2 C_1 — C_2 C_1
Contact point	3 rotations 2 translations	5	 C_1 C_2
Free joint	3 rotations 3 translations	6	No symbol No contact between bodies

Figure 11. ***Representation of mechanical articulations***

arms illustrated in Figure 9, the following description would be obtained: group 1, 3R; group 2, P-2R; group 3, 2R-P; group 4, P-R-P; group 5, 3P. This description is not sufficient to compute (eg in the system illustrated in Figure 12) the position of point A in relation to point O. It is not known about which axes the rotations and translations occur, nor to which axes they are related.

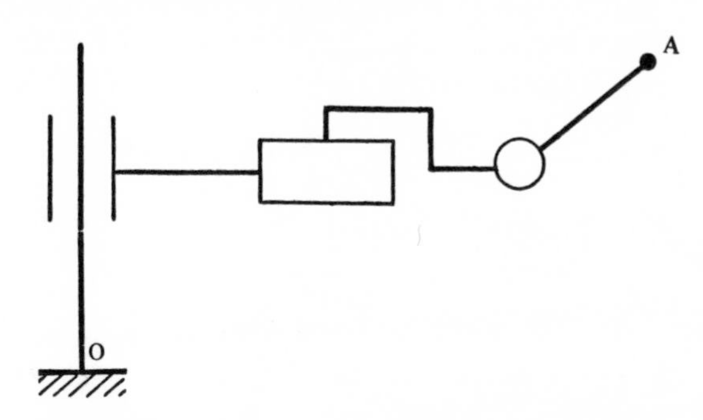

Figure 12. *An articulated system of group PR-PR or P-2R-P in the same order as the DOF*

For this reason Roth and Pieper complete their system of description with a number of *constant* or *variable* factors.

These factors can be defined with the help of Figure 13, which represents a manipulator,group 2R-P-R (after Borrel).

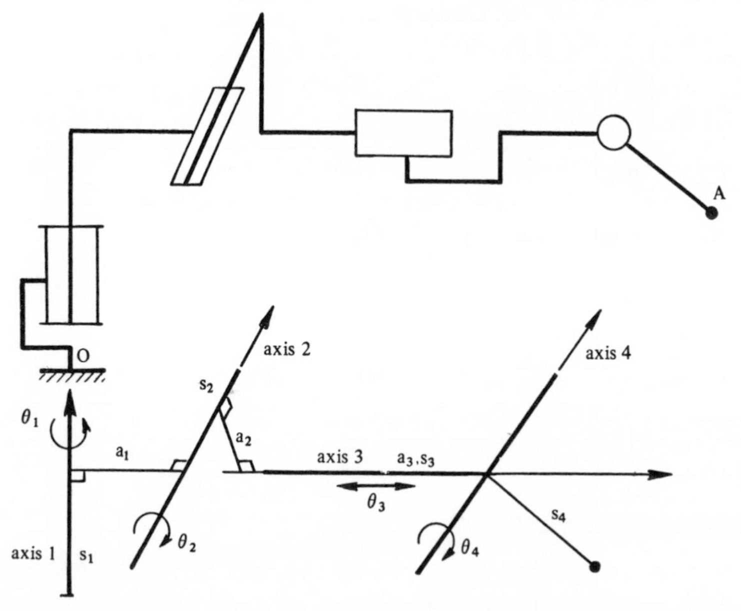

Figure 13. *System group 2R-P-R: factors used in the Roth-Pieper description*

Start by numbering and setting the orientation of the axes around which, and related to which, the articulated movements take place. The ith and $(i + 1)^{th}$ axes have a common normal which defines the distance a_i (for $i = 1, 2, 3$). The points of intersection of a_i and a_{i+1} with axis $i + 1$ define distance s_i measured in a positive sense of axis $i + 1$. The angle α_i, is formed by axes i and i+1, the positive sense being determined by the 'screw rule' applied to the axis $i + 1$ towards axis i. Then the rotation of a_i around axis i, to bring it to a_{i-1}, can be called θ_i.

With reference to these definitions it can be seen that:

(a) if the joint is rotational R, a_i, s_i and α_i are constants and θ_i is a variable.
(b) if the joint is translational P, a_i, α_i are constants and s_i is a variable.

With the system illustrated in Figure 13, the description would be:

$$2R\text{-}P\text{-}R\,(s_1 a_1 \alpha_1 \theta_1, s_2 a_2 \alpha_2 \theta_2, s_3 a_3 \alpha_3 \theta_3, s_4 a_4 \alpha_4 \theta_4)$$

θ_1, θ_2, θ_3 and θ_4 being variables. But for the last limb (axis 4) a_4, s_4 and α_4 are not defined since axis 5 does not exist. This is why Roth and Pieper added the following rule:

- If the last articulation is rotational, the variables are not included in the description.
- If the last articulation (axis n) is translational, write s_n.

For the example given, the following is obtained:

$$2R\text{-}P\text{-}R\,(s_1, a_1, \frac{3\pi}{2}, \theta_1; s_2, a_2, \frac{\pi}{2}, \theta_2; s_3, 0, \frac{\pi}{2}, \frac{\pi}{2})$$

One of the objectives of this description is to classify articulated systems. That is why in identifying the group it is possible to omit:

(a) indication of rotations and translations θ_1, θ_2 and s_3, since they are included in the preliminary formula (here 2R-P-R).
(b) all the angles that are multiples of $\pi/2$.

The manipulator illustrated in Figure 13 can be described by:

$$2R\text{-}P\text{-}R\,(s_1 a_1 s_2 a_2)$$

This method of description enables geometric and kinematic computation of the structure, but it does not allow development of dynamic models, lacking as it does the essential factors of mass, inertia, etc.

2.7.3 KHALIL'S DESCRIPTION[9]

This description has been deliberately constructed to allow the use of a computer, adapted to the generation of dynamic models of articulated systems; for this reason it includes the following elements:

1. The number of articulated segments constituting the system – NN.
2. A vector NDF (with a dimension NN) which gives the number of DOF acting on every segment.
3. A matrix MRR (with a dimension NN, 4) which specifies the nature and direction of the movement of each segment in relation to the preceding one, every row of this matrix corresponding to a DOF. In the last part of each row a rotation is denoted by 0, and a translation by 1. Of the three other terms in the row one, and only one, is not zero. The position of this term corresponds to the axis about which, or along which, the rotation or translation is accomplished (relative to a trihedral related to each articulation). The value of this element is determined by the number of DOF.
4. The vectors $L\lambda (1 \leqslant \lambda \leqslant NN)$ which define the set of Cartesian coordinates related to segment λ, the position of the origin of the set of coordinate axes related to segment $\lambda + 1$.
5. For each segment, the mass, position of the centre of mass and moment of inertia are given with reference to the set of coordinate axes related to the segment.

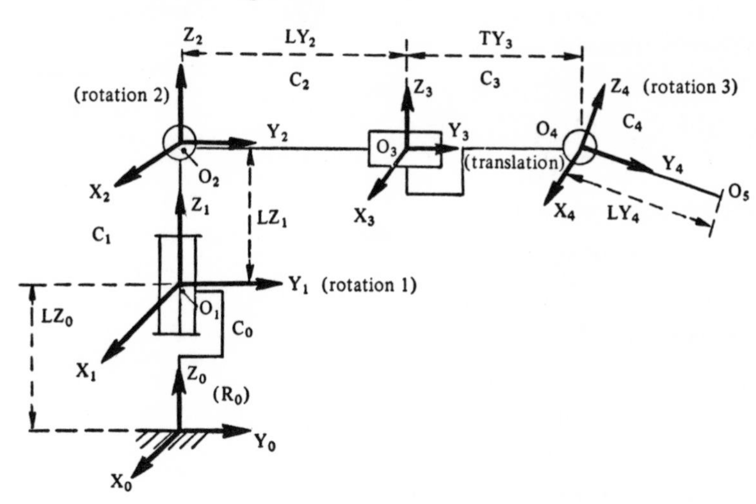

Figure 14. *The set of coordinate axes bound to the articulated system used in Khalil's description*

As an example, consider the articulated system illustrated in Figure 14. It has four segments, C_1 to C_4, which are mobile in relation to the segment C_0. For such a system, Khalil's description (inertial and mass elements excepted) can be written as:

NN = 4
NDF = 1, 2, 3, 4

$$MRR = \begin{pmatrix} 0 & 0 & 1 & 0 \\ 2 & 0 & 0 & 0 \\ 0 & 3 & 0 & 1 \\ 4 & 0 & 0 & 0 \end{pmatrix} \begin{matrix} \leftarrow C_1 \\ \leftarrow C_2 \\ \leftarrow C_3 \\ \leftarrow C_4 \end{matrix}$$

(columns: (x) (y) (z) (R or T))

$L^1 = (0, 0, LZ_1)^T$
$L^2 = (0, LY_2, 0)^T$
$L^3 = (0, TY_3, 0)^T$
$L^4 = (0, LY_4, 0)^T$

The description is easy to write and its formulation is suitable for computation, particularly in modelling. However, it cannot easily be used to compare two articulated structures.

2.7.4 THE RENAUD-ZABALA DESCRIPTION[10]

The aims of this description are similar to those of Khalil's. However, the description is symbolic, and it is not possible to give numerical values to the variables concerned. The procedure is as follows.

Initially a binary matrix [E] is generated, of which the elements are defined by $E_{ij} = 1$, if the segment i is linked directly to the segment j and is not situated between segments 0 and i. Otherwise, $E_{ij} = 0$. From the system illustrated in Figure 14 it is possible to obtain:

[E] =

i \ j	C_1	C_2	C_3	C_4
C_1	0	0	0	0
C_2	1	0	0	0
C_3	0	1	0	0
C_4	0	0	1	0

The matrix of relative movements (σ) is described thus:

$\sigma_i = 0$ if the ith segment pivots about an axis related to the preceding segment.
$\sigma_i = 1$ if the ith segment translates along an axis related to the preceding segment.

In the present example:

$$\begin{matrix} & C_1 & C_2 & C_3 & C_4 \\ [\sigma] = [& 0 & 0 & 1 & 0\ ]^T \end{matrix}$$

The Cartesian coordinate axes coupled with each segment are slightly different from those used by Khalil. The Z axis of each segment must coincide with the axis of movement of the segment relative to the one preceding it. Figure 15 shows the coordinate axes of the system illustrated in Figure 14.

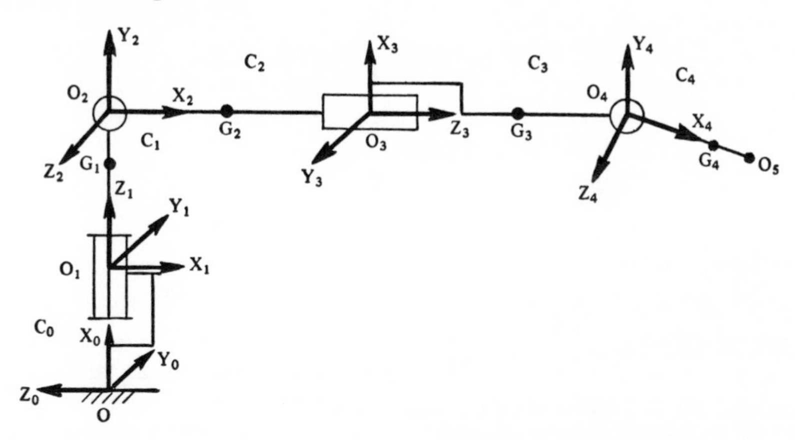

Figure 15. *The set of coordinate axes used in the Renaud-Zambala description*

The relative positions of the coordinate axes are a part of the description; they are defined by the vectors D (i, i + 1) (see later), which give the position of the origin of coordinate set i + 1 in coordinate set i. Also included are the vectors D (i, i), which give the position of the centre of mass of segment i in its own set of coordinate axes. Each element of these vectors is written as 0 if it is zero, 1 if it is constant, and 2 if it is variable. In the present example:

D (0,1) = (1,0,0)	D (1,1) = (0,0,1)
D (1,2) = (0,0,1)	D (2,2) = (1,0,0)
D (2,3) = (1,0,0)	D (3,3) = (0,0,2)
D (3,4) = (0,0,2)	D (4,4) = (1,0,0)

The relative orientations of the various sets of coordinate axes can be defined using Euler angles, α (i), β (i), γ (i) define the orientation of the coordinate set linked to segment i as compared with the coordinate set linked to segment i − 1. Only the components which are not zero and constant are given and for the articulated system illustrated in Figure 15 these angles are expressed in the following way:

$$\beta(1) = -\pi/2; \alpha(2) = -\pi/2; \alpha(3) = \gamma(3) = -\pi/2$$

The description, which like the previous one can be expressed in computational terms, is completed by the masses and moments of inertia of each segment.

2.7.5 BORREL'S DESCRIPTION[7]

Unlike other descriptions, which use data such as vectors and matrices, Borrel adhered to a linguistic description and wrote a compiler capable of interpreting this language. The description is appropriate to any segment of an articulated mechanism (without mechanical loops) in which each component is activated, relative to its neighbours, by rotational or translational movement. One of the components, C_0 (the base), can be placed in an orthogonal set of coordinates (R_0). A set of orthogonal coordinates bound to a terminal component is called R_{t1}.

2.7.5.1 Association between a graph and a mechanism

Borrel associated a directed graph with a mechanism. In this, the nodes are the intermediate coordinates between R_0 and R_{t1}, and the connecting arcs represent transfer from one coordinate set to another. Figure 16 shows an example of the graph-mechanism association. Only two types of coordinate axes are used:

(a) rotation of a segment about one of the axes of the preceding segment.
(b) translation of a segment along one of the axes of the preceding segment.

This implies that a transition through several intermediate coordinate sets is necessary in order to pass from the set of one segment to that of its predecessor.

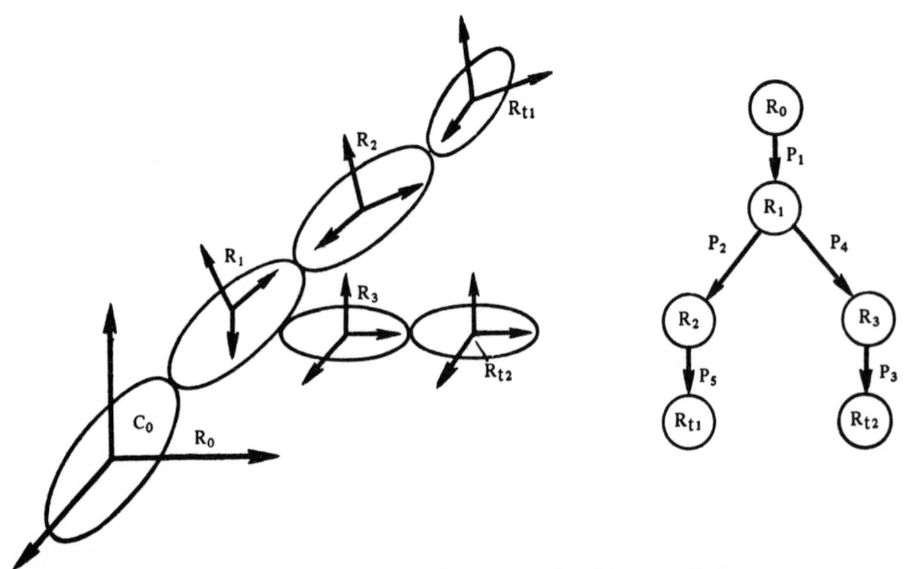

Figure 16. *An example of graph-mechanism association*

2.7.5.2 Definition of a chain

A chain can be defined as the part of a mechanism corresponding to a graph for which all the nodes except the first and the last are non-divergent. A chain is a succession of coordinate transformations associated with the arcs of the graph. Then:

$$HK = P_1 P_2 \ldots P_j \ldots P_n$$

if the chain is described by n transformations.

2.7.5.3 Description of a mechanism

A description of a mechanism can be obtained from the description of its constituent submechanisms. In turn, a submechanism can be described by the association with a subgraph, issuing from a divergent node. A mechanism can be represented by the graph illustrated in Figure 17.

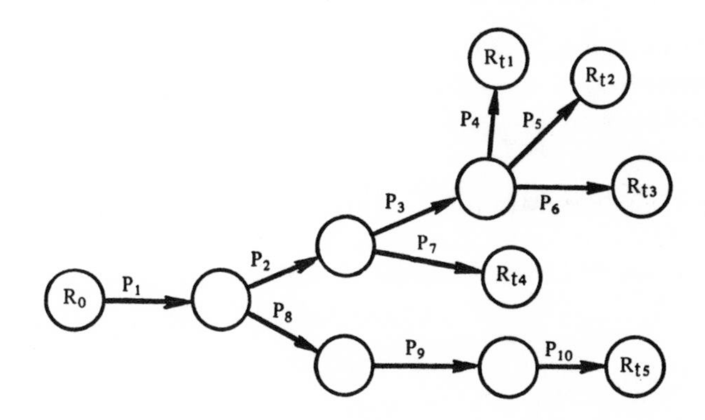

Figure 17. *A graph of a mechanism: the associated chains are $H1 = P_1$ $H2 = P_2$ $H3 = P_3$ $H4 = P_4$; $H5 = P_5$; $H6 = P_6$; $H7 = P_7$; $H8 = P_8, P_9, P_{10}$. The semicolons indicate the end of the mechanism; that is to say:*

$$M = H1 \left[[H8] \left[H2 [[H7] H3 [[H4][H5][H6]]] \right] \right]$$

$$M = P_1 \left[[P_8, P_9, P_{10};] \left[P_2 [P_7, P_3 [P_4; P_5; P_6;]] \right] \right]$$

2.7.5.4 *Notation for a coordinate transformation*

For use in a computer, a geometrical transformation may be described by:

- type: all character strings starting with R (rotation) or T (translation).
- axes: X, Y or Z.
- amplitude: if it is variable its maximum and minimum volumes should be noted in parentheses separated by a comma; if it is constant its value, preceded by the letter K, should be noted.

These values can be described numerically or alphanumerically. For rotation, the convention is to state the starting limit first and the range is then read in a clockwise direction.

2.7.5.5 *Description of a manipulator*

With reference to Figure 14, it can be supposed that the system has only one chain, and that the limits of variation of the articulated variables are:

ROT1 {R1 MIN = – 90; R1 MAX = 90} ROT2 {R2 MIN = – 30; R2 MAX = 50}

TR {T MIN = 0; T MAX = 150} ROT3 {R3 MI = – 20; R3 MA = 90}

and that the lengths have the values:

name	L1 (=OO_1)	L2 (=O_1O_2)	L3 (=O_2O_3)	L4 (=O_4O_5)
value	100	50	50	40

The system can then be described as:

TZK (L1), RZ(R1MIN, R1MAX), TZK(L2), RX(R2MIN, R2MAX), TYK(L3), TY(TMIN, TMAX), RX(R3MI, R3MA), TYK(L4);

or the equivalent expression:

TZK(100), RZ(−90, 90), TZK(50), RX(−30, 50), TYK(50), TY(0, 150), RX(−20, 90), TYK(40).

The simplicity of this description and its resemblance to natural language should be noted. However, it is not unique; a rotation can, for example, be written as a product of two other rotations. Further, all descriptions are linked to the choice of name of the axes of movement.

Conclusions

A robot is essentially a *functional* machine. Since the DOF are not equivalent, it is necessary to classify robots in the form of a hierarchy, in terms of the DOF of the vehicle, the arm and the end effector. The latter is that part of a robot which physically performs the task and, the longer the arm, the greater the effective movement at the extremity. This means that execution of a task is dependent on the DOF. This factor has not been discussed in any previous treatment of robotics. It should be taken into account in programs in which such descriptions are used, eg in computer modelling and simulation. Such descriptions could be a principal factor in the development of computer-aided design of robots.

Chapter 3

Articulated mechanical systems: determination of kinematic elements

In Chapter 2 it was shown that there are two principal classes of factors to be considered in an AMS:

- *structural* parameters, which are constant and determine the characteristics and organization of the constituent parts.
- *articulated* variables, which allow a mechanism to change its geometry; they are directly associated with the actuators and with the controls of the system.

The use of the AMS is related to its behaviour in real space (the task space). Real space can be represented by a set of Cartesian coordinates bound to the basic body, C_0, of a robot, having one of its axes coincident with the vertical.

Given the structural parameters, the kinematic elements of a robot are concerned with the determination of coordinates and the velocities of the different points of the articulated system, as well as the orientation and velocity of displacement of the various sets of coordinate axes linked to the body in the task space. These must be expressed as a function of the articulated variables (often called *generalized variables*).

The kinematic elements play a prime role in the construction of every AMS and it is important, therefore, to study their computation. A classical method can be used to do this which employs three-dimensional vectors in real space. Other authors, such as Paul,[11] have used a four-dimensional space to make their equations more applicable.

A vector can be represented by its three orthogonal components and by a scale factor which is associated with their effectiveness. The vector is then unchanged if its four components are multiplied by another single scalar factor.

3.1 Computation of the orientation of a chain relative to a set bound to an upper segment

Take an AMS comprised of a chain of segments (see Borrel's description); to each of these segments, C_0, C_i, . . . C_n, is bound a

Cartesian group, R_0 (X_0 Y_0 Z_0) . . . R_i (X_i Y_i Z_i) . . . R_n (X_n Y_n Z_n).

To pass from group R_i to group R_{i+1} (Figure 18) there can be as many as six geometrical transformations, ie three rotations about the axes X : Y : Z and three translations along these axes.

To simplify the calculation, it can be supposed that to pass from group R_i to group R_{i+1} only one of these six transformations is possible. This leads (going from R_i to R_{i+1}) to the addition of, at most, five simulated segments (Figure 19), possessing neither mass nor size. The role of these added segments is to ensure that the previous condition (only one transformation between one segment and another) is upheld.

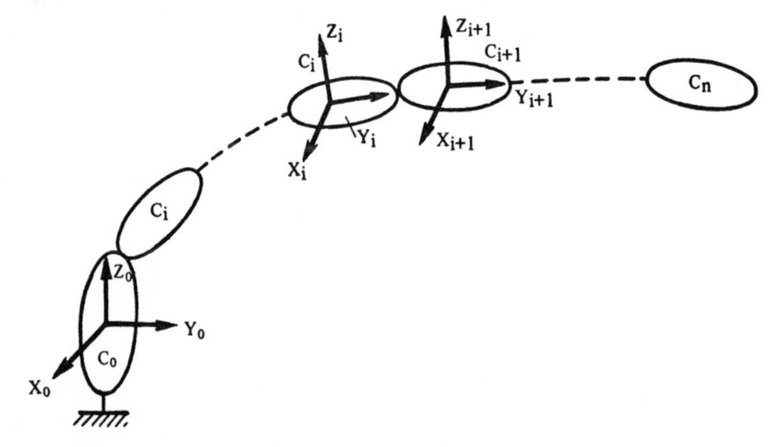

Figure 18. *An articulated chain and its associated set of coordinates*

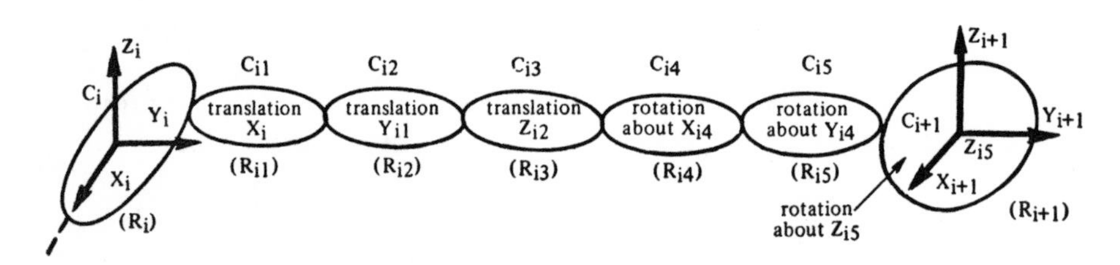

Figure 19. *The insertion of simulated segments C_{ij} between C_i and C_{i+1}*

The main interest here, however, is the change of *orientation* of the coordinate axes and it is evident that every translation can be represented by a matrix unity coordinate transformation. The imaginary segments are, at most, sufficient between C_i and C_{i+1}.

Consider a complete chain of all the real and simulated segments, numbered from C_0 to C_n. To change from R_i to R_{i+1}, suppose that there is a rotation θ_k of R_i about X_i. Then:

and

$$\begin{pmatrix} X_i \\ Y_i \\ Z_i \end{pmatrix} = \begin{pmatrix} 1 & 1 & 1 \\ 0 & \cos\theta_k & -\sin\theta_k \\ 0 & \sin\theta_k & \cos\theta_k \end{pmatrix} \begin{pmatrix} X_{i+1} \\ Y_{i+1} \\ Z_{i+1} \end{pmatrix} \qquad (3\text{-}1)$$

$$\underline{V}\,(R_i) = M_i^{i+1}\,\underline{V}\,(R_{i+1}) \qquad (3\text{-}2)$$

Given a vector $\underline{V}$ in R_{i+1}, its components in set R_i can be calculated by using transformation (3-2). If the rotation θ_k is made about Y_i it can be written as:

$$M_i^{i+1} = \begin{pmatrix} \cos\theta_k & 0 & \sin\theta_k \\ 0 & 1 & 0 \\ -\sin\theta_k & 0 & \cos\theta_k \end{pmatrix} \qquad (3\text{-}3)$$

If the rotation θ_k is made about Z_i the transformation becomes:

$$M_i^{i+1} = \begin{pmatrix} \cos\theta_k & -\sin\theta_k & 0 \\ \sin\theta_k & \cos\theta_k & 0 \\ 0 & 0 & 1 \end{pmatrix} \qquad (3\text{-}4)$$

Equation (3-2) is recurrent; for instance:

$$\underline{V}\,(R_{i-1}) = M_{i-1}^{i}\,.\,M_i^{i+1}\,\underline{V}\,(R_{i+1}) \qquad (3\text{-}5)$$

Then:

$$M_{i-1}^{i+1} = M_{i-1}^{i}\,.\,M_i^{i+1} \qquad (3\text{-}6)$$

and:

$$\underline{V}\,(R_p) = M_p^{p+\lambda}\,\underline{V}\,(R_{p+\lambda}) \; (\lambda \geqslant 0) \qquad (3\text{-}7)$$

$$M_p^{p+\lambda} = M_p^{p+1}\,.\,M_{p+1}^{p+2} \ldots M_{p+\lambda-2}^{p+\lambda-1}\,.\,M_{p+\lambda-1}^{p+\lambda} \qquad (3\text{-}8)$$

Equations (3-7) and (3-8) describe the orientation of $R_{p+\lambda}$ (the axial unity vector) in R_p. It is then possible to calculate the coordinate sets related to each of the segments, relative to the reference coordinates, related to the base C_0.

Example 1: Calculation of the unity vectors of the set of coordinate axes linked to the gripper illustrated in Figure 20.[12]

The different transformations of the set of coordinate axes are:

M_0^1 : rotation about Z_0 through θ_1

$$M_0^1 = \begin{pmatrix} C\theta_1 & -S\theta_1 & 0 \\ S\theta_1 & C\theta_1 & 0 \\ 0 & 0 & 1 \end{pmatrix} \qquad C\theta_1 \equiv \cos\theta_1, \quad S\theta_1 \equiv \sin\theta_1$$

M_1^2 : rotation about X_1 through θ_2

$$M_1^2 = \begin{pmatrix} 1 & 0 & 0 \\ 0 & C\theta_2 & S\theta_2 \\ 0 & S\theta_2 & C\theta_2 \end{pmatrix}$$

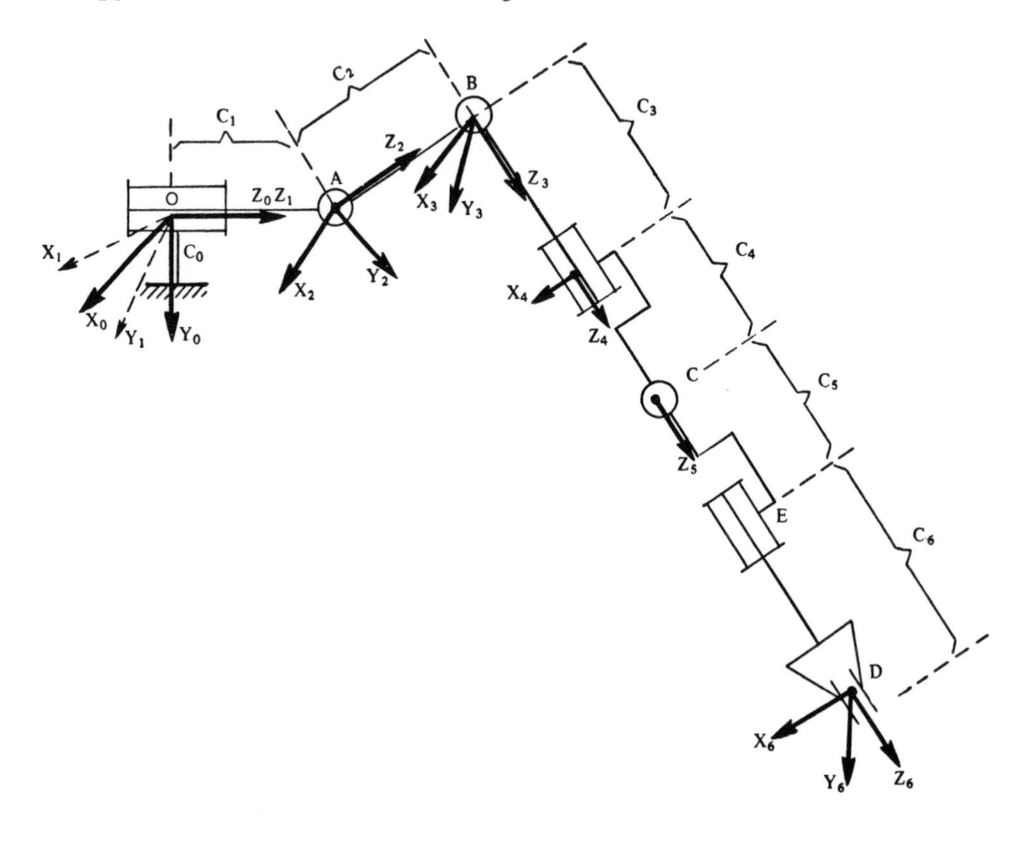

Figure 20. *The structure of a manipulator and associated set of coordinate axes*

M_2^3 : rotation about X_2 through θ_3

$$M_2^3 = \begin{pmatrix} 1 & 0 & 0 \\ 0 & C\theta_3 & -S\theta_3 \\ 0 & S\theta_3 & C\theta_3 \end{pmatrix}$$

M_3^4 : rotation about Z_3 through θ_4

$$M_3^4 = \begin{pmatrix} C\theta_4 & -S\theta_4 & 0 \\ S\theta_4 & C\theta_4 & 0 \\ 0 & 0 & 1 \end{pmatrix}$$

M_4^5 : rotation about Z_4 through θ_5

$$M_4^5 = \begin{pmatrix} 1 & 0 & 0 \\ 0 & C\theta_5 & -S\theta_5 \\ 0 & S\theta_5 & C0_5 \end{pmatrix}$$

M_5^6 : rotation about Z_5 through θ_6

$$M_5^6 = \begin{pmatrix} C\theta_6 & -S\theta_6 & 0 \\ S\theta_6 & C\theta_6 & 0 \\ 0 & 0 & 1 \end{pmatrix}$$

Using equations (3-7) and (3-8):

$$\underline{V}(R_0) = M_0^6 \ \underline{V}(R_6)$$

$$M_0^6 = M_0^1 \ M_1^2 \ M_2^3 \ M_3^4 \ M_4^5 \ M_5^6$$

The unity vectors of the set of coordinate axes of the gripper are as follows. To simplify the transcription:

$$CI = \cos \theta_i \qquad (3\text{-}9)$$

$$SI = \sin \theta_i \qquad (3\text{-}10)$$

$$C\,(I + J) = \cos (\theta_i + \theta_j) \qquad (3\text{-}11)$$

$$S\,(I + J) = \sin (\theta_i + \theta_j) \qquad (3\text{-}12)$$

which gives the result:

$$\begin{aligned}\underline{x_6} = [&C1C4C6 - S1C\,(2{+}3)\,S4C6 - C1S4C5S6 \\ &- S1C\,(2{+}3)\,C4C5S6 + S1S\,(2{+}3)\,S5S6]\,\underline{x_0} \\ &+ [S1C4C6 + C1C\,(2{+}3)\,S4C6 - S1S4C5S6 \\ &+ C1C\,(2{+}3)\,C4C5S6 - C1S\,(2{+}3)\,S5S6]\,\underline{y_0} \\ &+ [S(2{+}3)S4C6 + S(2{+}3)C4C5S6 + C(2{+}3)S5S6]\,\underline{z_0}\end{aligned}$$

$$\begin{aligned}\underline{y_6} = &[-C1C4S6 + S1C(2{+}3)S4S6 - C1S4C5C6 - S1C(2{+}3)C4C5C6 + S1S(2{+}3)S5C6]\,\underline{x_0} \\ &+ [-S1C4S6 - C1C(2{+}3)S4S6 - S1S4C5C6 + C1C(2{+}3)C4C5C6 - C1S(2{+}3)S5C6]\,\underline{y_0} \\ &+ [-S(2{+}3)S4S6 + S(2{+}3)C4C5C6 + C(2{+}3)S5C6]\,\underline{z_0}\end{aligned}$$

$$\begin{aligned}\underline{z_6} = &[C1SS4S5 + S1C(2{+}3)C4S5 + S1S(2{+}3)C5]\,\underline{x_0} \\ &+ [S1S4S5 - C1C(2{+}3)C4S5 - C15(2{+}3)C5]\,\underline{y_0} \\ &+ [-S(2{+}3)C4S5 + C(2{+}3)C5]\,\underline{z_0}\end{aligned}$$

Example 2: Calculation of the unity vectors in R_4 for the mechanism (containing a translation) illustrated in Figure 14.
The transformations of the coordinate axes are:

M_0^1 : rotation, angle θ_1 about Z_0

M_1^2 : rotation, angle θ_2 about X_1

M_2^3 : translation following Y_2

M_3^4 : rotation, angle θ_3 about X_3.

Using the notations described earlier:

$$M_0^1 = \begin{pmatrix} C1 & -S1 & 0 \\ S1 & C1 & 0 \\ 0 & 0 & 1 \end{pmatrix} \quad M_1^2 = \begin{pmatrix} 1 & 0 & 0 \\ 0 & C2 & -S2 \\ 0 & S2 & C2 \end{pmatrix} \quad M_2^3 = \begin{pmatrix} 1 & 0 & 0 \\ 0 & 1 & 0 \\ 0 & 0 & 1 \end{pmatrix} \quad M_3^4 = \begin{pmatrix} 1 & 0 & 0 \\ 0 & C3 & -S3 \\ 0 & S3 & C3 \end{pmatrix}$$

$$M_0^4 = M_0^1 \ M_1^2 \ M_2^3 \ M_3^4 = M_0^1 \ M_1^2 \ M_3^4$$

which gives:

$$\begin{pmatrix} \underline{x_4} \\ \underline{y_4} \\ \underline{z_4} \end{pmatrix} = \begin{pmatrix} C1 & -S1C2C3+S1S2S3 & S1C2S3+S1S2C3 \\ S1 & C1C2C3 - C1S2S3 & -C1C2S3-C1S2C3 \\ 0 & S2C3+C2S3 & -S2S3+C2C3 \end{pmatrix} \begin{pmatrix} \underline{x_0} \\ \underline{y_0} \\ \underline{z_0} \end{pmatrix} \qquad (3\text{-}13)$$

3.2 Computation of the orientation of a chain relative to a set bound to a lower segment

The matrices M_i^{i+1}, which represent a rotation equal or not equal to zero (translation), has the following characteristic:

$$[M_i^{i+1}]^{-1} = [M_i^{i+1}]^T \tag{3-14}$$

The inverse of M_i^{i+1} is M_{i+1}^i, which corresponds to the translation from R_i to R_{i+1}:

$$M_{i+1}^i = [M_i^{i+1}]^T \tag{3-15}$$

The use of this recurrence gives (as described in section 3.1):

$$\begin{aligned} M_{p+\lambda}^p &= [M_p^{p+\lambda}]^{-1} = [M_p^{p+1} \cdot M_{p+1}^{p+2} \ldots M_{p+\lambda-2}^{p+\lambda-1} \cdot M_{p+\lambda-1}^{p+\lambda}]^{-1} \\ &= [M_{p+\lambda-1}^{p+\lambda}]^{-1} [M_{p+\lambda-2}^{p+\lambda-1}]^{-1} \ldots [M_{p+1}^{p+2}]^{-1} [M_p^{p+1}]^{-1} \\ &= [M_{p+\lambda-1}^{p+\lambda}]^T [M_{p+\lambda-2}^{p+\lambda-1}]^T \ldots [M_{p+1}^{p+2}]^T [M_p^{p+1}]^T \end{aligned} \tag{3-16}$$

Equation (3-16) gives, by a simple transposition of the matrix $M_p^{p+\lambda}$, the orientation of the coordinate set of axes of solid C_p relative to that of solid $C_{p+\lambda}$. For example, consider the orientation of the coordinate set, R_0, with respect to R_4, which is associated with the gripper (see Figure 14):

$$M_4^0 = (M_3^4)^T (M_1^2)^T (M_0^1)^T = (M_0^4)^T \tag{3-17}$$

and then refer to equation (3-13):

$$\begin{pmatrix} \underline{x_0} \\ \underline{y_0} \\ \underline{z_0} \end{pmatrix} = \begin{pmatrix} C1 & S1 & 0 \\ -S1C2C3+S1S2S3 & C1C2C3-C1S2S3 & S2C3+C2S3 \\ S1C2S3+S1S2C3 & -C1C2S3-C1S2C3 & -S2C3+C2C3 \end{pmatrix} \begin{pmatrix} \underline{x_4} \\ \underline{y_4} \\ \underline{z_4} \end{pmatrix}$$

3.3 Computation of the position of a point on a chain in relation to an upper segment

It can be supposed that corresponding to each component, C_q, of a chain there is a corresponding coordinate set, R_q, with an origin O_q (Figure 21).

To calculate the position of point Q in the component C_λ in the set of coordinate axes related to component $C_{\lambda-p}$ ($p \leqslant \lambda$), the following vectorial relationship can be written:

$$\underline{O_{\lambda-p} Q} = \underline{O_{\lambda-p} O_\lambda} + \underline{O_\lambda Q} \tag{3-18}$$

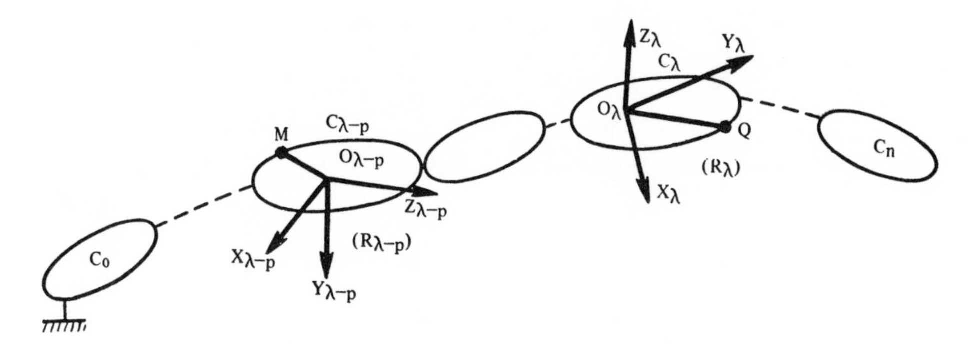

Figure 21a. *An articulated chain and associated set of coordinate axes*

If this relationship is expressed in set $R_{\lambda-p}$, the following is obtained:

$$\underline{O_{\lambda-p}Q}\,(R_{\lambda-p}) = \underline{O_{\lambda-p}\,O_\lambda}\,(R_{\lambda-p}) + \underline{O_\lambda Q}\,(R_{\lambda-p}) \qquad \textbf{(3-19)}$$

The vector $\underline{O_\lambda Q}$, appertaining to segment C_λ in set R_λ, is known. To determine its value in set $R_{\lambda-p}$, it is necessary to apply the previously defined transformation, $M^\lambda_{\lambda-p}$. In set $R_{\lambda-p}$ this provides the components of a vector determined in set R_λ. Equation (3-19) can then be written as:

$$\underline{O_{\lambda-p}\,Q}\,(R_{\lambda-p}) = \underline{O_{\lambda-p}O_\lambda}\,(R_{\lambda-p}) + M^\lambda_{\lambda-p}\cdot\underline{O_\lambda Q}\,(R_\lambda) \qquad \textbf{(3-20)}$$

with

$$M^\lambda_{\lambda-p} = M^{\lambda-p+1}_{\lambda-p}\;\ldots\;M^{\lambda-1}_{\lambda-2}\;M^\lambda_{\lambda-1} \qquad \textbf{(3-21)}$$

Equation (3-20) is recurrent and allows calculation of the coordinates of any point on a chain, in relation to a preceding segment.

For example, take the manipulator illustrated in Figure 20; calculating the coordinates of D in the set appertaining to segment C_0, using the notations $\ell_1 = OA$, $\ell_2 = AB$, $\ell_3 = BC$ and $\ell_4 = CD$:

$$\underline{OD}\,(R_0) = \underline{OC}\,(R_0) + M^5_0\,\underline{CD}\,(R_5)$$
$$\underline{OC}\,(R_0) = \underline{OB}\,(R_0) + M^4_0\,\underline{BC}\,(R_4)$$
$$\underline{OB}\,(R_0) = \underline{OA}\,(R_0) + M^2_0\,\underline{AB}\,(R_2)$$

and

$$\underline{OA}\,(R_0) = \underline{OA}\,(R_1) = [0, 0, \ell_1]^T$$
$$\underline{AB}\,(R_2) = [0, 0, \ell_2]^T$$
$$\underline{BC}\,(R_4) = [0, 0, \ell_3]^T$$
$$\underline{CD}\,(R_5) = [0, 0, \ell_4]^T = \underline{CD}\,(R_6)$$

Therefore:

$$\underline{OD}\,(R_0) = \begin{pmatrix}0\\0\\\ell_1\end{pmatrix} + M^1_0\,M^2_1\begin{pmatrix}0\\0\\\ell_2\end{pmatrix} + M^1_0\,M^2_1\,M^3_2\,M^4_3\begin{pmatrix}0\\0\\\ell_3\end{pmatrix} + M^1_0\,M^2_1\,M^3_2\,M^4_3\,M^5_4\begin{pmatrix}0\\0\\\ell_4\end{pmatrix}$$

The matrices M_i^{i+1} can be determined using the procedures described in section 3.1 and the following results can be obtained:

$$X_D(R_0) = S1S2\ell_2 + (S1C2S3+S1S2C3)\ell_3 + [S5(S4C1-C4(S1S2S3-S1C2C3)) + C5(S1C2S3+S1S2C3)]\ell_4$$

$$Y_D(R_0) = -C1S2\ell_2 - (C1C2S3+C1S2C3)\ell_3 + [S5(S1S4-C4C1(C2C3-S2S3)) - C5C1(C2S3+S2C3)]\ell_4$$

$$Z_D(R_0) = \ell_1 + C2\ell_2 + (C2C3-S2S3)\ell_3 + [C5C(2+3)-C4S5S(2+3)]\ell_4$$

As a further example, consider the mechanism illustrated in Figure 14 containing a translation; calculation of the coordinates of O_5 in set R_0 gives:

$$\underline{O\,O_5}\,(R_0) = \underline{O\,O_4}\,(R_0) + M_0^4\,\underline{O_4O_5}\,(R_4)$$
$$\underline{O\,O_4}\,(R_0) = \underline{O\,O_3}\,(R_0) + M_0^3\,\underline{O_3O_4}\,(R_3)$$
$$\underline{O\,O_3}\,(R_0) = \underline{O\,O_2}\,(R_0) + M_0^2\,\underline{O_2O_3}\,(R_2)$$
$$\underline{O\,O_2}\,(R_0) = [\,0, 0, LZ_0 + LZ_1\,]^T$$
$$\underline{O_2O_3}\,(R_2) = [\,0, LY_2, 0\,]^T$$
$$\underline{O_3O_4}\,(R_3) = [\,0, TY_3, 0\,]^T$$
$$\underline{O_4O_5}\,(R_4) = [\,0, LY_4, 0\,]^T$$

and the matrices M_i^{i+1} have been given in the previous paragraph:

$$X_{O_5}(R_0) = -(TY_3 + LY_2)\,S1C2 + LY_4\,(S1S2S3 - S1C2C3)$$
$$Y_{O_5}(R_0) = (TY_3 + LY_2)\,C1C2 + LY_4\,(C1C2C3 - C1S2S3)$$
$$Z_{O_5}(R_0) = LZ_0 + LZ_1 + (LY_2 + TY_3)\,S2 + LY_4\,(S2C3 + C2S3)$$

3.4 Computation of the position of a point on a chain in relation to a lower segment

With reference to Figure 21, the following can be written:

$$\underline{O_\lambda M}\,(R_\lambda) = \underline{O_\lambda\,O_{\lambda-p}}(R_\lambda) + M_\lambda^{\lambda-p}\,.\,\underline{O_{\lambda-p}\,M}\,(R_{\lambda-p})$$

and, knowing that according to equation (3-16):

$$M_\lambda^{\lambda-p} = [M_{\lambda-p}^\lambda]^T$$

calculation of the position of a point is simple. However, in practice, it is better to assign sets of coordinate axes to each segment (whatever be the relative positions of the sets on the chain), starting from the lowest segment, which is assumed to be fixed.

3.5 Determination of the velocity vectors of rotation of different segments of a chain relative to a set of coordinate axes

The movement of set R_λ, related to segment C_λ, relative to another segment of a chain (for example $C_{\lambda-p}$), can be expressed kinematically by a velocity vector which is the sum of the instantaneous translational and rotational speeds. The instantaneous velocity vector $(\underline{\Omega}_\lambda^{\lambda-p})$ of coordinate set R_λ related to segment C_λ (in relation to $R_{\lambda-p}$, related to the segment $C_{\lambda-p}$), must be expressed in a particular coordinate set. For example, if $R_{\lambda-p}$ is chosen, the following can be written:

$$\underline{\Omega}_\lambda^{(\lambda-p)}\ (R_{\lambda-p})$$

Theory of the mechanics[13] allows calculation of this vector (assumed to be known in set $R_{\lambda-p}$) in set R_λ, by using an intermediate set of coordinate axes, for example R_q, and matrices related to the set transformation (M_i^j), as described earlier (see Figure 22):

$$\underline{\Omega}_\lambda^{(\lambda-p)}\,(R_\lambda) = \overset{+}{\underline{\Omega}}{}_\lambda^{(\lambda-1)}\,(R_{\lambda-1}) + M_\lambda^q\ \underline{\Omega}_q^{(\lambda-p)}\,(R_q) \qquad \textbf{(3-21b)}$$

The term $\overset{+}{\underline{\Omega}}{}_\lambda^{(\lambda-1)}$ $(R_{\lambda-1})$ represents the rotation of set R_λ relative to a set with axes parallel to those in set $R_{\lambda-1}$ but with an origin O_λ (see Figure 21b).

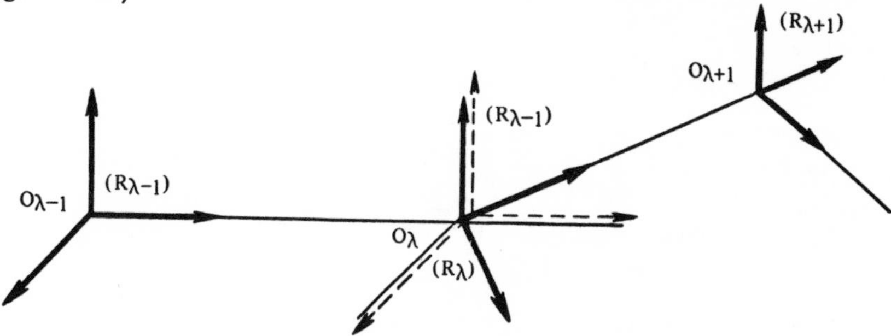

Figure 21b. *Definition of* $\overset{+}{\underline{\Omega}}{}_\lambda^{(\lambda-1)}$ $(R_{\lambda-1})$

Figure 22. *A chain and associated set of coordinate axes: the rotational vectors do not have their origin bound to the set of coordinate axes*

If, as is generally the case, segment C_0 (the fixed base of the chain) is taken as reference and, as an intermediate set, that linked to segment C_q (so that $q = \lambda - 1$), a recurrent relationship is established which is useful in these computations — and is particularly useful in the evaluation of the kinematic energy of a dynamic model:

$$\underline{\Omega}_\lambda^{(0)}(R_\lambda) = \overset{+}{\underline{\Omega}}{}_\lambda^{(\lambda-1)}(R_{\lambda-1}) + M_\lambda^{\lambda-1}\,\underline{\Omega}_{\lambda-1}^{(0)}(R_{\lambda-1}) \qquad (3\text{-}22)$$

or:

$$\underline{\Omega}_\lambda^{(0)}(R_\lambda) = \overset{+}{\underline{\Omega}}{}_\lambda^{(\lambda-1)}(R_{\lambda-1}) + [M_{\lambda-1}^{\lambda}]^T\,\underline{\Omega}_{\lambda-1}^{(0)}(R_{\lambda-1}) \qquad (3\text{-}23)$$

Example: Calculation of the instantaneous velocity of the sets linked with the parts of the manipulator arm illustrated in Figure 20.
The following origin and speed vectors are known:

$$\underline{\Omega}_0^{(0)}(R_0) = [\,0,\ 0,\ 0\,]^T;\quad \overset{+}{\underline{\Omega}}{}_1^{(0)}(R_0) = [\,0,\ 0,\ \dot\theta_1\,]^T;\quad \overset{+}{\underline{\Omega}}{}_2^{(1)}(R_1) = [\dot\theta_2,\ 0,\ 0\,]^T;$$

$$\overset{+}{\underline{\Omega}}{}_3^{(2)}(R_2) = [\,\dot\theta_3,\ 0,\ 0\,]^T;\quad \overset{+}{\underline{\Omega}}{}_4^{(3)}(R_3) = [\,0,\ 0,\ \dot\theta_4\,]^T;\quad \overset{+}{\underline{\Omega}}{}_5^{(4)}(R_4) = [\,\dot\theta_5,\ 0,\ 0\,]^T;$$

$$\overset{+}{\underline{\Omega}}{}_6^{(5)}(R_5) = [\,0,\ 0,\ \dot\theta_6\,]^T$$

Using equation (3-23), the following can be written:

$$\underline{\Omega}_1^{(0)}(R_1) = \overset{+}{\underline{\Omega}}{}_1^{(0)}(R_0) + [M_0^1]^T\,\underline{\Omega}_0^{(0)}(R_0)$$

$$\underline{\Omega}_1^{(0)}(R_1) = [\,0,\ 0,\ \dot\theta_1\,]^T$$

$$\underline{\Omega}_2^{(0)}(R_2) = \overset{+}{\underline{\Omega}}{}_2^{(1)}(R_1) + [M_1^2]^T\,\underline{\Omega}_1^{(0)}(R_1)$$

$$= \begin{pmatrix}\dot\theta_2\\0\\0\end{pmatrix} + \begin{pmatrix}1&0&0\\0&C2&S2\\0&-S2&C2\end{pmatrix}\begin{pmatrix}0\\0\\\dot\theta_1\end{pmatrix} = [\dot\theta_2,\ S2\,\dot\theta_1,\ C2\,\dot\theta_1\,]^T$$

$$\underline{\Omega}_3^{(0)}(R_3) = \overset{+}{\underline{\Omega}}{}_3^{(2)}(R_2) + [M_2^3]^T\,\underline{\Omega}_2^{(0)}(R_2)$$

$$= \begin{pmatrix}\dot\theta_3\\0\\0\end{pmatrix} + \begin{pmatrix}1&0&0\\0&C3&S3\\0&-S3&C3\end{pmatrix}\begin{pmatrix}\dot\theta_2\\S2\,\dot\theta_1\\C2\,\dot\theta_1\end{pmatrix} = [\dot\theta_2 + \dot\theta_3,\ \dot\theta_1\,S(2+3),\ \dot\theta_1\,C(2+3)]^T$$

etc.

$$\underline{\Omega}_6^{(0)}(R_6) = \begin{pmatrix}\dot\theta_5 + (\dot\theta_2 + \dot\theta_3)C4 + \dot\theta_1\ S4S(2+3)\\ \dot\theta_4\,S5 + \dot\theta_1\ S5C(2+3) + \dot\theta_1\,C4C5S(2+3) - C5S4(\dot\theta_2 + \dot\theta_3)\\ \dot\theta_4\,C5 + \dot\theta_1\ C5C(2+3) - \dot\theta_1\,C4S5S(2+3) + S4S5(\dot\theta_2 + \dot\theta_3) + \dot\theta_6\end{pmatrix}$$

3.6 Determination of the velocity vectors of translation of different segments of a chain relative to a set of coordinate axes

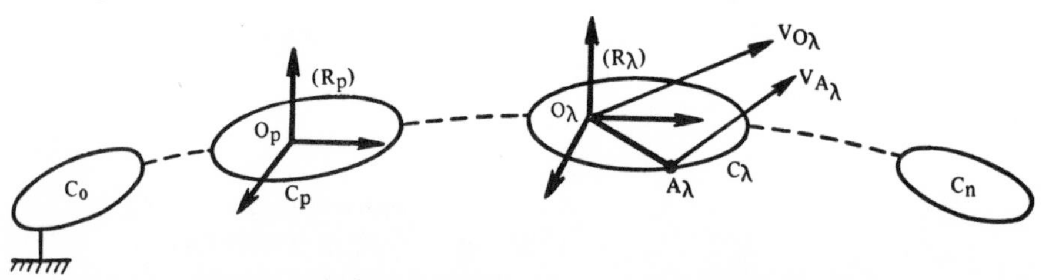

Figure 23. *A chain and associated set of coordinate axes*

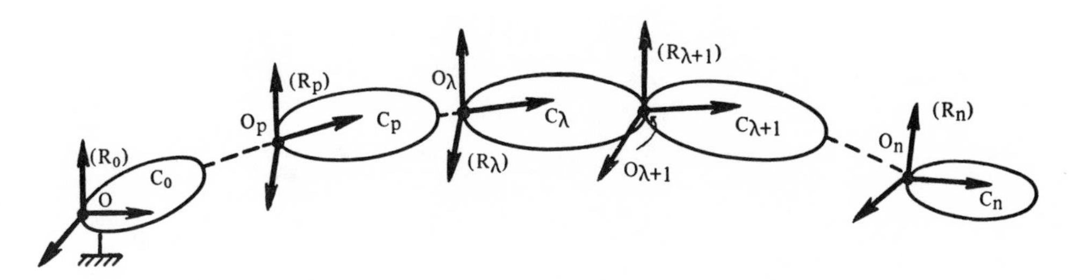

Figure 24.

Let $\underline{V}^{(p)}_{A\lambda}(R_\lambda)$ be the velocity of the translational vector of point A_λ appertaining to segment C_λ, relative to the set linked to segment C_p; this vector can be expressed in set R_λ. Using simple laws of mechanics, the following expression may be written:

$$\underline{V}^{(p)}_{A_\lambda}(R_\lambda) = \underline{V}^{(p)}_{O_\lambda}(R_\lambda) + \underline{\Omega}^{(p)}_{\lambda}(R_\lambda) \wedge \underline{O_\lambda A_\lambda}(R_\lambda) \qquad \textbf{(3-24)}$$

If point A_λ is chosen as being common to segments C_λ and $C_{\lambda+1}$, to which the origin of the coordinate of $C_{\lambda+1}$ is fixed (Figure 24), it is possible to obtain from equation (3-24):

$$\underline{V}^{(p)}_{O_{\lambda+1}}(R_\lambda) = \underline{V}^{(p)}_{O_\lambda}(R_\lambda) + \underline{\Omega}^{(p)}_{\lambda}(R_\lambda) \wedge \underline{O_\lambda O_{\lambda+1}}(R_\lambda) \qquad \textbf{(3-25)}$$

If the set R_λ, related to the body C_0, is taken as reference, the following relationship can be obtained:

$$\underline{V}^{(0)}_{O_{\lambda+1}}(R_\lambda) = \underline{V}^{(0)}_{O_\lambda}(R_\lambda) + \underline{\Omega}^{(0)}_{\lambda}(R_\lambda) \wedge \underline{O_\lambda O_{\lambda+1}}(R_\lambda) \qquad \textbf{(3-26)}$$

with:

$$\underline{V}^{(0)}_{O_{\lambda+1}}(R_\lambda) = M^{\lambda+1}_{\lambda}\ \underline{V}^{(0)}_{O_{\lambda+1}}(R_{\lambda+1}) \qquad \textbf{(3-27)}$$

Equation (3-27) is identical to the relationship described by equation (3-2) which links components of the same vector in two neighbouring coordinate sets.

Example: Calculation of the translational speed of points A, B, C and E of the manipulator arm illustrated in Figure 20.
Here $OA = \ell_1$, $AB = \ell_2$, $BC = \ell_3$, and $CE = \ell_4$ relative to set R_0 linked to segment C_0.

$$\underline{V}^{(0)}_{O}(R_0) = [\,0,\ 0,\ 0\,]^T$$

$$\underline{V}^{(0)}_{A}(R_1) = [\,0,\ 0,\ 0\,]^T$$

$$\underline{V}^{(0)}_{B}(R_2) = \underline{V}^{(0)}_{A}(R_2) + \underline{\Omega}^{(0)}_{2}(R_2) \wedge \underline{AB}(R_2)$$

$$= \begin{pmatrix}0\\0\\0\end{pmatrix} + \begin{pmatrix}\dot\theta_2\\ \dot\theta_1\,S2\\ \dot\theta_1\,C2\end{pmatrix} \wedge \begin{pmatrix}0\\0\\ \ell_2\end{pmatrix}$$

$$= [\ell_2\,\dot\theta_1\,S2,\ -\ell_2\,\dot\theta_2,\ 0\,]^T$$

$$\underline{V}_C^{(0)}(R_3) = \underline{V}_B^{(0)}(R_3) + \underline{\Omega}_3^{(0)}(R_3) \wedge \underline{BC}(R_3)$$

$$= [M_2^3]^T \underline{V}_B^{(0)}(R_2) + \underline{\Omega}_3^{(0)}(R_3) \wedge \underline{BC}(R_3)$$

$$= \begin{pmatrix} 1 & 0 & 0 \\ 0 & C3 & S3 \\ 0 & -S3 & C3 \end{pmatrix} \begin{pmatrix} \ell_2 \dot\theta_1 S2 \\ -\ell_2 \dot\theta_2 \\ 0 \end{pmatrix} + \left(\begin{pmatrix} \dot\theta_2 + \dot\theta_3 \\ \dot\theta_1 S(2+3) \\ \dot\theta_1 C(2+3) \end{pmatrix} \wedge \begin{pmatrix} 0 \\ 0 \\ \ell_3 \end{pmatrix} \right)$$

$$\underline{V}_C^{(0)}(R_3) = [\ell_2 \dot\theta_1 S2 + \ell_2 \dot\theta_1 S(2+3), -\ell_2 \dot\theta_2 C3 - \ell_3 (\dot\theta_2 + \dot\theta_3), \ell_2 \dot\theta_2 S3]^T$$

$$\underline{V}_C^{(0)}(R_4) = [M_3^4]^T \underline{V}_C^{(0)}(R_3)$$

$$\underline{V}_C^{(0)}(R_4) = \begin{pmatrix} C4 \dot\theta_1 (\ell_2 S2 + \ell_3 S(2+3)) + S4 (\ell_2 \dot\theta_2 C3 + \ell_3 (\dot\theta_2 + \dot\theta_3)) \\ S4 \dot\theta_1 (\ell_2 S2 + \ell_3 S(2+3)) - C4 (\ell_2 \dot\theta_2 C3 + \ell_3 (\dot\theta_2 + \dot\theta_3)) \\ \ell_2 \dot\theta_2 S_3 \end{pmatrix}$$

$$\underline{V}_E^{(0)}(R_5) = \underline{V}_C^{(0)}(R_5) + \underline{\Omega}_5^{(0)}(R_5) \wedge \underline{CE}(R_5)$$

$$\underline{V}_C^{(0)}(R_5) = [M_4^5]^T V_C^{(0)}(R_4)$$

$$\underline{V}_E^{(0)}(R_5) = \begin{pmatrix} C4[\ell_2\dot\theta_1 S2 + \ell_3\dot\theta_1 S(2+3)] - S4[\ell_2\dot\theta_2 C3 + \ell_3(\dot\theta_2 + \dot\theta_3)] + [\ell_4 \\ (S5\dot\theta_4 - S4C5(\dot\theta_2 + \dot\theta_3) + \dot\theta_1 S(2+3)C4C5 + \dot\theta_1 S5C(2+3))] \\ -S4C5[\ell_2\dot\theta_1 S2 + \ell_3\dot\theta_1 S(2+3)] - C4C5[\ell_2\dot\theta_2 C3 + \ell_3(\dot\theta_2 + \dot\theta_3)] + \\ \ell_2 S3S5\dot\theta_2 - \ell_4[\dot\theta_5 + C4(\dot\theta_2 + \dot\theta_3) + \dot\theta_1 S4S(2+3)] \\ S4S5[\ell_2\dot\theta_1 S2 + \ell_3\dot\theta_1 S(2+3)] + C4S5[\ell_2\dot\theta_2 C3 + \ell_3(\dot\theta_2 + \dot\theta_3)] \\ + \ell_2 S3C5\dot\theta_2 \end{pmatrix}$$

3.6.2 IN AN ARTICULATED CHAIN CONTAINING TRANSLATIONS

In the articulated chain illustrated in Figure 25 the chain undergoes a translation between segments C_λ and $C_{\lambda+1}$ and the following can be written:

$$\underline{V}_{O_{\lambda+1}}^{(p)}(R_\lambda) = \underline{V}_{O_\lambda}^{(p)}(R_\lambda) + \underline{\Omega}_\lambda^{(p)}(R_\lambda) \wedge O_\lambda O_{\lambda+1}(R_\lambda) + \underline{V}_{O_{\lambda+1}}^{O_\lambda}(R_\lambda) \quad (3\text{-}28)$$

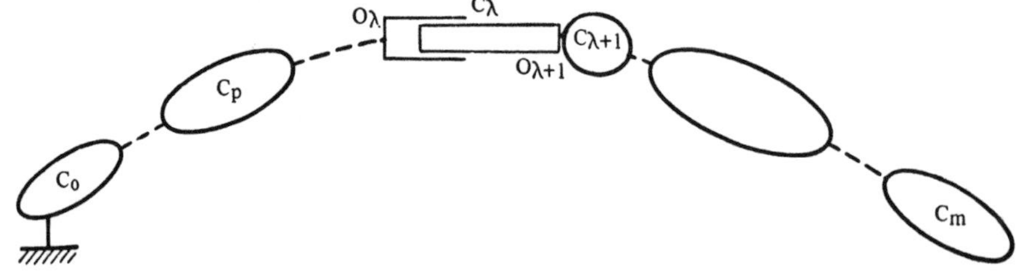

Figure 25. *An articulated chain with a single translation*

Referring to the lower body ℓ_0, (p) can be replaced with (0). Equation (3-28) is similar to equation (3-26) but in the former the translational speed of $O_{\lambda+1}$ relative to O_λ is added.

Example: Calculation of the relative velocity of translation of points O_1, O_2, O_3, O_4 and O_5 on the manipulator arm illustrated in Figure 14.

$$\underline{\Omega}_0^{(0)}(R_0) = [0,0,0]^T \quad \underline{\Omega}_1^{(0)}(R_1) = [0,\ 0,\ \dot\theta_1]^T$$

$$\underline{\Omega}_2^{(1)}(R_1) = [\dot\theta_2, 0,\ 0]^T \quad \underline{\Omega}_3^{(2)}(R_2) = [0,\ 0,\ 0]^T$$

$$\underline{\Omega}_4^{(3)}(R_3) = [\dot\theta_3, 0,\ 0]^T$$

$$\underline{V}_{O_1}^{0}(R_0) = \underline{V}_{O_2}^{O_1}(R_1) = \underline{V}_{O_3}^{O_2}(R_2) = \underline{V}_{O_5}^{O_4}(R_4) = [0,\ 0,\ 0]^T$$

$$\underline{V}_{O_4}^{O_3}(R_3) = [0,\ \dot T,\ 0]^T$$

These expressions provide data which allow the required values to be calculated. These are:

$$\rightarrow \underline{V}_{O_2}^{(0)}(R_2) = [0,\ 0,\ 0]^T$$

$$\underline{V}_{O_3}^{(0)}(R_2) = \underline{V}_{O_2}^{(0)}(R_2) + \underline{\Omega}_2^{(0)}(R_2) \wedge \underline{O_2O_3}(R_2)$$

$$\underline{\Omega}_2^{(0)}(R_2) = \overset{+}{\underline{\Omega}}{}_2^{(1)}(R_1) + [M_1^2]^T\ \underline{\Omega}_1^{(0)}(R_1)$$

$$= \begin{pmatrix} \dot\theta_2 \\ 0 \\ 0 \end{pmatrix} + \begin{pmatrix} 1 & 0 & 0 \\ 0 & C2 & S2 \\ 0 & -S2 & C2 \end{pmatrix} \begin{pmatrix} 0 \\ 0 \\ \dot\theta_1 \end{pmatrix} = \begin{pmatrix} \dot\theta_2 \\ S2\,\dot\theta_1 \\ C2\,\dot\theta_1 \end{pmatrix}$$

$$\underline{V}_{O_3}^{(0)}(R_2) = [LY_2.C2.\dot\theta_1,\ 0,\ -LY_2.\dot\theta_2]^T$$

$$\rightarrow \underline{V}_{O_3}^{(0)}(R_3) = [M_2^3]^T\ \underline{V}_{O_3}^{(0)}(R_2) = \underline{V}_{O_3}^{(0)}(R_2)$$

$$\underline{V}_{O_4}^{(0)}(R_3) = \underline{V}_{O_3}^{(0)}(R_3) + \underline{\Omega}_3^{(0)}(R_3) \wedge \underline{O_3O_4}(R_3) + \underline{V}_{O_4}^{O_3}(R_3)$$

$$\underline{\Omega}_3^{(0)}(R_3) = \overset{+}{\underline{\Omega}}{}_3^{(2)}(R_2) + [M_2^3]^T\ \underline{\Omega}_2^{(0)}(R_2) = \underline{\Omega}_2^{(0)}(R_2)$$

$$\underline{V}_{O_4}^{(0)}(R_3) = \begin{pmatrix} LY_2.C2\dot\theta_1 \\ 0 \\ -LY_2.\dot\theta_2 \end{pmatrix} + \begin{pmatrix} \dot\theta_2 \\ S2\,\dot\theta_1 \\ C2\,\dot\theta_1 \end{pmatrix} \wedge \begin{pmatrix} 0 \\ TY_3 \\ 0 \end{pmatrix} + \begin{pmatrix} 0 \\ \dot{TY}_3 \\ 0 \end{pmatrix}$$

$$= [C2.\dot\theta_1\ TY_3 + LY_2.C2.\dot\theta_1,\ \dot{TY}_3,\ -\dot\theta_2\ TY_3 - LY_2.\dot\theta_2]^T$$

$$\underline{V}_{O_4}^{(0)}(R_4) = [M_3^4]^T\ \underline{V}_{O_4}^{(0)}(R_3) = \begin{pmatrix} 1 & 0 & 0 \\ 0 & C3 & S3 \\ 0 & -S3 & C3 \end{pmatrix} \underline{V}_{O_4}^{(0)}(R_3)$$

$$\rightarrow \underline{V}_{O_4}^{(0)}(R_4) = \begin{pmatrix} C2\,\dot\theta_1\,(TY_3+LY_2) \\ C3.\dot{TY}_3 - S3\,\dot\theta_2\,(TY_3+LY_2) \\ -S3.\dot{TY}_3 - C3\,\dot\theta_2\,(TY_3+LY_2) \end{pmatrix}$$

$$\underline{V}^{(0)}_{O_5}(R_4) = \underline{V}^{(0)}_{O_4}(R_4) + \underline{\Omega}^{(0)}_4(R_4) \wedge \underline{O_4O_5}(R_4)$$

$$\underline{\Omega}^{(0)}_4(R_4) = \overset{+}{\underline{\Omega}}{}^{(3)}_4(R_3) + [M_3^4]^T\, \underline{\Omega}^{(0)}_3(R_3)$$

$$= \begin{pmatrix} \dot{\theta}_3 \\ 0 \\ 0 \end{pmatrix} + \begin{pmatrix} 1 & 0 & 0 \\ 0 & C3 & S3 \\ 0 & -S3 & C3 \end{pmatrix} \begin{pmatrix} \dot{\theta}_2 \\ S2\,\dot{\theta}_1 \\ C2\,\dot{\theta}_1 \end{pmatrix} = \begin{pmatrix} \dot{\theta}_2 + \dot{\theta}_3 \\ C3\,S2\,\dot{\theta}_1 + S3C2\dot{\theta}_1 \\ -S3S2\,\dot{\theta}_1 + C3\,C2\,\dot{\theta}_1 \end{pmatrix}$$

$$\rightarrow \underline{V}^{(0)}_{O_5}(R_4) = \begin{pmatrix} \dot{\theta}_1\,[C2\,(TY_3 + LY_2) + LY_4\,C(2+3)] \\ C3\,.\,\dot{TY}_3 - S3\,.\,\dot{\theta}_2\,(TY_3 + LY_2) \\ -S3\,.\,TY_3 - C3\,(TY_3 + LY_2)\,\dot{\theta}_2 - LY_4\,(\dot{\theta}_2 + \dot{\theta}_3) \end{pmatrix}$$

Conclusions

A knowledge of the possible orientations, positions and velocities of the component segments of an articulated mechanical chain is necessary before elaboration of the modelling required for computer control of such a chain is possible. Starting from a treatment of the structural parameters and coordinates of the segments, the kinematic elements can be calculated with ease, using the relationships described in this chapter. It should be noted that these recurrent relationships can be used in programming, but that the results become complicated with increasing numbers of segments and changes of coordinates. These preliminary relationships will now permit an investigation into the problems of control of articulated robots.

Chapter 4

Calculation of robot articulation variables

The problem to be studied in this chapter is crucial to the use of an AMS. Suppose that the robot illustrated in Figure 26 is to be used to grip an object, using a structure such as the manipulator shown in Figure 20. Knowing the location of the table and the cylinder, the required position of the robot must be determined. Then, with the manipulator in configuration 1, it is necessary to determine the changes which need to be applied to the various articulated variables to bring them to configuration 2 (whilst avoiding obstacle A). These problems involve all aspects of robot control: plan and trajectory generation and the reaction with the environment. A necessary step in solving the problem of control is to establish transformation of known data concerning the task space to data concerning the articulated variables.

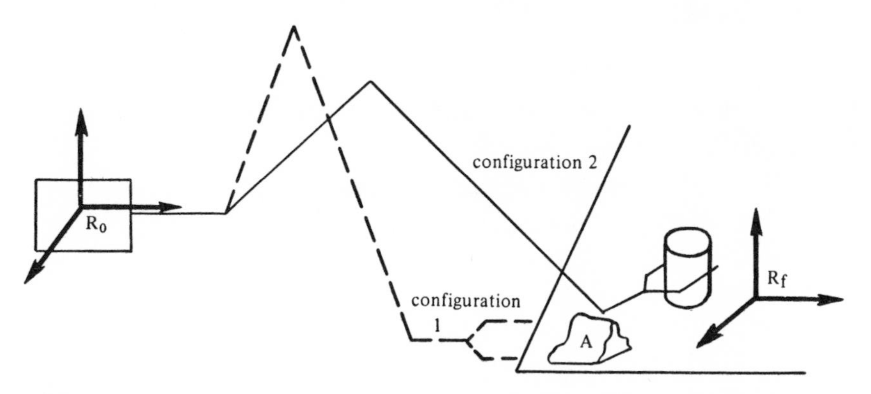

Figure 26. *The grasping of an object using a manipulator having six DOF*

Solutions for the variables of an articulated robot are defined by the answers given to the following inference equation:

knowledge of the structural variables of the robot (and of their limits) (= a description of the robot)	+	knowledge of points (P_i) in the task space and/or of directions (S_j) of interest	+	knowledge of the spatial constraints in which the task is performed	→	calculation of the values of the articulated variables

This is a static equation which, in fact, describes the problem of coordinate transformation between the task space (three-dimensional) and the articulated variable space (six-dimensional at most). Put another way, given set (P_i, S_j) in the task space, what value must be given to $\underline{\Theta} = (\theta_1, \theta_2 \ldots \theta_n)$, n being the number of DOF?

This problem may not have a solution, or may have a limited number of solutions, or an infinite number of solutions. Each of these possibilities will now be examined.

4.1 The absence of a solution

There are three possible causes for the absence of a solution – geometrical, mechanical and mathematical.

Geometrical

The set (P_i, S_j), relating to data or constraints of the task space [defined by the coordinate set R_0: $P_i(R_0)$, $S_j(R_0)$], is incompatible with or contradictory to the geometry of the mechanism.

Example: Consider the mechanism with two DOF shown in Figure 27, (R_0) being its task space.
Imposing point P_0:

$$P_2(R_0): X_{O_2} = L_1; Y_{O_2} = 2L_2$$

it is clear that this point cannot be reached by the mechanism if $L_2 > 2/3\ L_1$. In the same way, imposing:

$$P_1(R_0)\ (X_{O_1} = L_1; Y_{O_1} = 0)$$
$$P_2(R_0)\ [X_{O_2} = L_1; Y_{O_2} = (L_1^2 + L_2^2)^{1/2}]$$

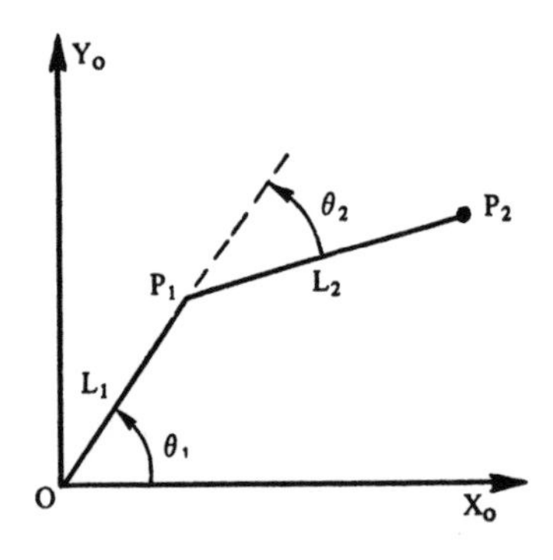

Figure 27.

the first constraint is satisfied for $\theta_1 = 0$, but the second is incompatible with the first. If there is incompatibility between the geometry of the mechanism and the constraints imposed by the values of $[P_i(R_0), S_j(R_0)]$ there is, in fact, no solution.

Mechanical

This is an extension of the geometrical problem. The following are not normally included when representing the geometry of a mechanism:

- range limits of each articulated variable (assuming each rotation to have a range of 360° and each translation to be unlimited).
- possible couplings between articulated variables; they arise from the nature of the transmission between the actuators and the articulations.

For example, in manipulator MA23 (see Figure 20) rotations θ_3 (about X_2) and θ_4 (about Z_3) are not independent for mechanical reasons – and similarly, θ_5 (about X_4) and θ_6 (about Z_5). This further limits the choice of the values of $[P_i(R_0), S_j(R_0)]$ which will enable the solution for articulated variables.

Mathematical

To illustrate this example, consider the robot shown in Figure 14. Suppose that $[P_i(R_0), S_j(R_0)]$ is reduced to a knowledge of point O_5 in R_0. In section 3.3 the following equations were established:

$$X_{O_5}(R_0) = (TY_3 + LY_2)S1C2 + LY_4(S1S2S3 - S1C2C3)$$
$$Y_{O_5}(R_0) = (TY_3 + LY_2)C1C2 + LY_4(C1C2C3 - C1S2S3)$$
$$Z_{O_5}(R_0) = LZ_0 + LZ_1 + (LY_2 + TY_3)S2 + LY_4(S2C3 + S3C2)$$

Suppose that O_5 can be solved both geometrically and mechanically; then the equations above must be solvable. Inverting the problem, and knowing that $X_{O_5}(R_0)$, $Y_{O_5}(R_0)$ and $Z_{O_5}(R_0)$ are compatible with the constraints, what are the values of θ_1, θ_2, TY_3 and θ_3?

In this case there are four unknown variables for three equations, which may indicate an infinity of solutions (see section 4.2). By defining θ_3 three equations are obtained with three unknown variables. But since these equations are non-linear, it would be difficult to find solutions for θ_1, θ_2 and TY_3. The greater the number of coupled equations, the more difficult their solution becomes. When using computer control this coordinate transformation must always be made. How this problem can be solved in practice, by an appropriate choice of $[P_i(R_0), S_j(R_0)]$, will be considered later.

4.2 An infinite number of solutions

When (Pi, Sj) is compatible with feasible configurations of the robot, but the number of constraints (K) (or components of Pi, Sj) is lower than the number of DOF (N), there are, as in the previous example, K equations having N unknowns ($K < N$). By analytical solutions $K - N$ articulated variables are obtained, of which the value could be arbitrary (whilst remaining compatible with the geometry of the mechanism).

Taking the example of the class 3 arm illustrated in Figure 28, in which the coordinates of C in R_0 are to be found using the methods described in Chapter 3, the following expressions can be written:

$$M_0^1 = \begin{pmatrix} C1 & -S1 & 0 \\ S1 & C1 & 0 \\ 0 & 0 & 1 \end{pmatrix} \quad M_1^2 = \begin{pmatrix} 1 & 0 & 0 \\ 0 & C2 & -S2 \\ 0 & S2 & C2 \end{pmatrix} \quad M_2^3 = (\Uparrow)$$

$$\underline{OC}(R_0) = \underline{OB}(R_0) + M_0^3\, \underline{BC}(R_3)$$

$$\underline{OB}(R_0) = \underline{OA}(R_0) + M_0^2\, \underline{AB}(R_2)$$

from which:

$$X_C(R_0) = -S1C2(L_2 + T)$$

$$Y_C(R_0) = C1C2(L_2 + T)$$

$$Z_C(R_0) = L_1 + S2(L_2 + T)$$

If $[P_i(R_0), S_j(R_0)]$ are defined by the constraints:

$$X_C = 0, Y_C = L_1 + L_2$$

the solutions to system (4-1) are given by:

$$\theta_1 = 0, C2(L_2 + T) = L_1 + L_2$$

or

$$\theta_1 = \pi, C2(L_2 + T) = -(L_1 + L_2)$$

A double infinity of combinations for θ_2 and T can then be obtained.

4.3 A limited number of solutions

If the constraints imposed by $[P_i(R_0), S_j(R_0)]$ are incompatible with the mechanical reality, and if their number is equal to the number of DOF, the mathematical solutions become limited in number. Because of the non-linearity of the equations this number is often >1. However, these solutions cannot always be computed and some are not mechanically feasible.

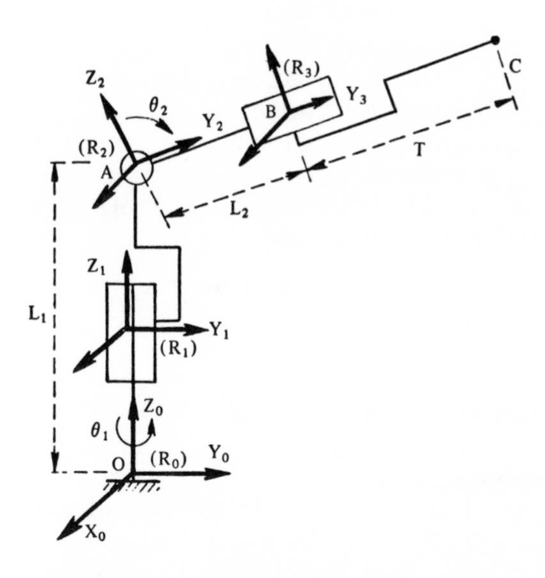

Figure 28. *An example of a class 3 arm (2R-P)*

Consider the previous example: if, in addition to $X_C\ (R_0) = 0$, $Y_C\ (R_0) = L_1 + L_2$, the constraint $Z_C\ (R_0) = L_1$ is imposed, then the following system can be written:

$$\begin{aligned} S1C2(L_2 + T) &= 0 \\ C1C2(L_2 + T) &= L_1 + L_2 \\ S2(L_2 + T) &= 0 \end{aligned} \qquad \textbf{(4-1)}$$

which gives the solution:

$$\begin{array}{lll} \theta_1 = 0 & \theta_2 = 0 & T = L_1 \\ \theta_1 = 0 & \theta_2 = \pi & T = L_1 - 2L_2 \\ \theta_1 = \pi & \theta_2 = 0 & T = L_1 - 2L_2 \\ \theta_1 = \pi & \theta_2 = \pi & T = L_1 \end{array}$$

If, for mechanical reasons, T can only be positive and θ_2 can vary only between 0 and $\pi/2$, then only one of the four mathematical solutions is feasible:

$$\theta_1 = \theta_2 = 0; \quad T = L_1$$

4.4 Practical choice of $[P_i(R_0), S_j(R_0)]$

The function of a robot is to bring about a change to its environment, using its end effector(s) (which might be grippers or tools). From the viewpoint of control, the operator is interested in:

1. Determination of the values of the articulated variables from data available on the orientation and position of the end effector.
2. The presence of obstacles which might impede the movement of the end effector: such events must be considered when deciding on the configuration to be used. In practice, however, this problem can generally be avoided (an exception being when robots are used on continuous production lines).

This first problem will be considered for a mechanism with a maximum of six DOF (see Chapter 2 for exceptions).

4.5 Mechanisms with six degrees of freedom

Robotic structures are typically comprised of an arm (with three DOF) which is used to change the position of the end effector in the task space. The end effector is normally capable of three rotations about three orthogonal axes. It is of particular interest that these three axes intersect because it then becomes possible to decouple the positioning of the base of the end effector from its orientation (very useful for purposes of control). However, in practice, this occurs only rarely (usually for reasons of design, eg size). Figure 29 shows an example in which the rotational axes do not converge. In Figure 30 the axes do converge and the structure is much improved.

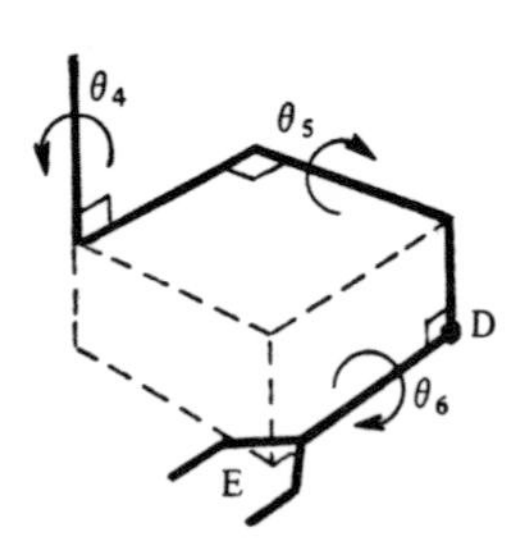

Figure 29. *The structure of an end effector in which the three rotational axes do not converge*

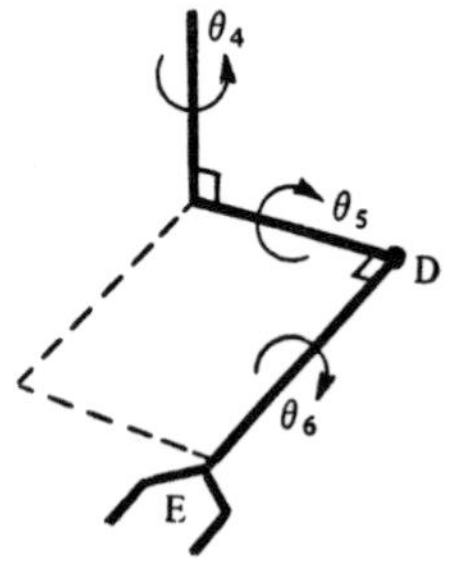

Figure 30. *The structure of an end effector with intersecting 2 x 2 axes*

Consider Figure 31 where the length of the end effector is known; in this case it is the position of D that is of interest. When the three axes intersect, the position of D is dependent only on the arm. In the systems illustrated in Figures 29 and 30, as the gripper is orientated the position of D changes (unless, that is, the gripper rotates). Thus, the position of D is dependent on the arm and on the angles θ_4 and θ_5.

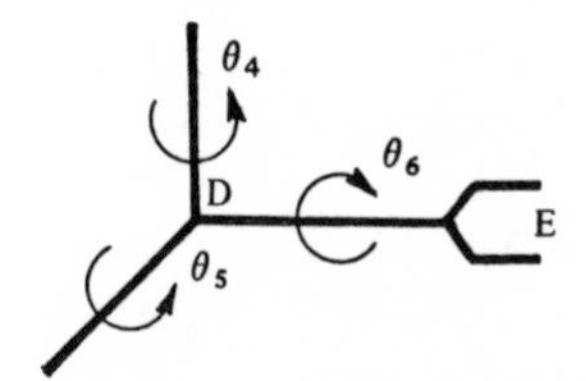

Figure 31. *An example of an end effector with three intersecting axes*

As an example, consider Figure 32, in which a gripper is attached by a swivel joint to the arm illustrated in Figure 28.

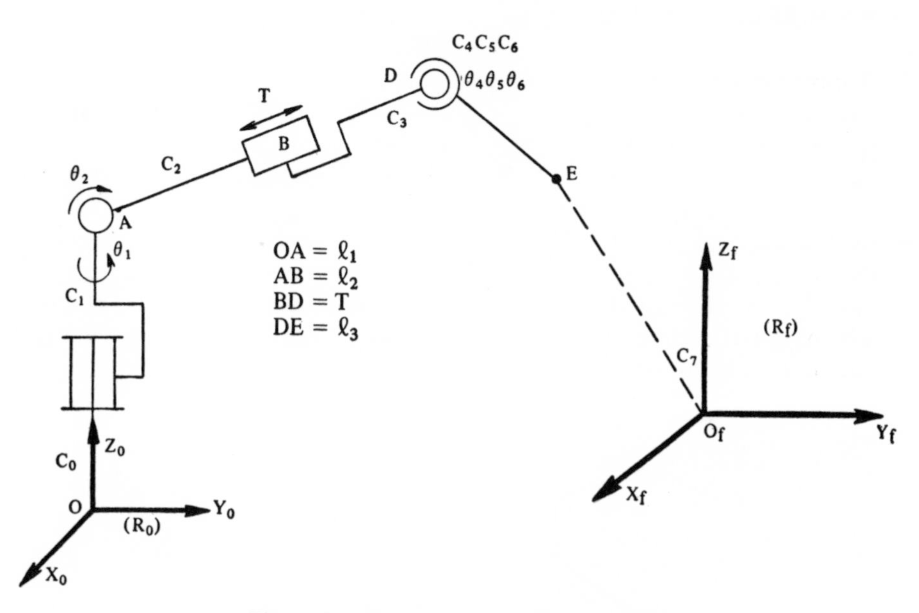

Figure 32. *A robot possessing six DOF*

The constraints which need to be considered here are the position of E and the direction ED relative to the task axes (R_f), or the position of D and the direction DE relative to R_f. For the first hypothesis the relevant data are:

$$[P_i(R_f), S_j(R_f)] = \begin{pmatrix} X_E(R_f) & \alpha_f \\ Y_E(R_f) & \beta_f \\ Z_E(R_f) & \gamma_f \end{pmatrix} \qquad (4\text{-}2)$$

where α_f, β_f and γ_f are the cosines directrix of ED in R_f and R_0 is known relative to R_f. But, how are θ_1, θ_2, T, θ_4, θ_5 and θ_6 calculated?

1. Determination of the coordinates of D in R_0: It is found that:

$$
\begin{aligned}
X_D(R_f) &= X_E(R_f) + \alpha_f \ell_3 \\
Y_D(R_f) &= Y_E(R_f) + \beta_f \ell_3 \\
Z_D(R_f) &= Z_E(R_f) + \gamma_f \ell_3
\end{aligned} \qquad (4\text{-}3)
$$

and then, according to equation (3-20), for R_0:

$$\underline{OD}(R_0) = \underline{OO_f}(R_0) + M_0^f \, \underline{O_fD}(R_f)$$

For instance, by simplifying R_0 and R_f, separated by a known translation, the following can be obtained:

$$
\begin{aligned}
[M_0^f] &= [\Uparrow] \\
X_D(R_0) &= X_{O_f}(R_0) + X_E(R_f) + \alpha_f \ell_3 \\
Y_D(R_0) &= Y_{O_f}(R_0) + Y_E(R_f) + \beta_f \ell_3 \\
Z_D(R_0) &= Z_{O_f}(R_0) + Z_E(R_f) + \gamma_f \ell_3
\end{aligned} \qquad (4\text{-}4)
$$

2. *Computation of θ_1, θ_2 and T:* Equation (4-1) provides the relationships:

$$
\left.
\begin{aligned}
X_D(R_0) &= -S1C2(\ell_2+T) \\
Y_D(R_0) &= C1C2(\ell_2+T) \\
Z_D(R_0) &= \ell_1+S2(\ell_2+T)
\end{aligned}
\right\} \qquad (4\text{-}5)
$$

with:

$$\theta_1^* = \arctan(-X_D / Y_D) \qquad (4\text{-}6)$$

$$\theta_2^* = \arctan[C\theta_1^* (Z_D - \ell_1)/Y_D] \qquad (4\text{-}7)$$

$$T_1^* = \ell_2 + [(Z_D - \ell_1)/S\theta_2^*] \qquad (4\text{-}8)$$

$$T_2^* = \ell_2 - [(Z_D - \ell_1)/S\theta_2^*] \qquad (4\text{-}9)$$

and equation (4-5) provides four possible solutions:

$$(\theta_1^*, \theta_2^*, T_1^*)\,;(\theta_1^*, \theta_2^* + \pi, T_2^*)\,;(\theta_1^* + \pi, -\theta_2^*, T_2^*)\,;(\theta_1^* + \pi, -\theta_2^* + \pi, T_1^*)$$

Some of these solutions must be eliminated because they are not mechanically feasible – this leaves a choice of two.

3. *Computation of θ_4, θ_5 and θ_6:* As was discussed in section 3.1, if three successive coordinate sets are associated with segment DE constituting segments C_4, C_5 and C_6, of which two are simulated, the following expression can be obtained:

$$
\begin{aligned}
\underline{DE}(R_6) &= [0, 0, \ell_3]^T = M_6^3 \, \underline{DE}(R_3) \\
\underline{DE}(R_3) &= \underline{EO_f}(R_3) + \underline{O_fO}(R_3) + \underline{OD}(R_3) \\
&= M_3^f \, [\underline{EO_f}(R_f) + \underline{O_fO}(R_f) + \underline{OD}(R_f)] \\
&= M_3^0 \, [\Uparrow] \, [\underline{EO_f}(R_f)] + M_3^0 \, [\Uparrow] \, [\underline{O_fO}(R_0) + \underline{OD}(R_0)]
\end{aligned}
$$

$$
\begin{pmatrix} 0 \\ 0 \\ \ell_3 \end{pmatrix} = M_6^3 \left[M_3^0 \begin{pmatrix} X_E(R_f) \\ Y_E(R_f) \\ Z_E(R_f) \end{pmatrix} + M_3^0 \begin{pmatrix} X_{O_f}(R_0) + X_D(R_0) \\ Y_{O_f}(R_0) + Y_D(R_0) \\ Z_{O_f}(R_0) + Z_D(R_0) \end{pmatrix} \right] \qquad (4\text{-}10)
$$

By writing:

$$\delta_x = X_E(R_f) + X_{O_f}(R_0) + X_D(R_0)$$
$$\delta_y = Y_E(R_f) + Y_{O_f}(R_0) + Y_D(R_0)$$
$$\delta_z = Z_E(R_f) + Z_{O_f}(R_0) + Z_D(R_0)$$

and then describing equation (4-10) in the form:

$$M_3^6 \begin{pmatrix} 0 \\ 0 \\ \ell_3 \end{pmatrix} = [M_0^3]^T \begin{pmatrix} \delta_x \\ \delta_y \\ \delta_z \end{pmatrix} \qquad \textbf{(4-11)}$$

equation (4-11) is obtained from which θ_4, θ_5 and θ_6 can be determined, since:

$$M_3^6 = \begin{pmatrix} 1 & 0 & 0 \\ 0 & C4 & -S4 \\ 0 & S4 & C4 \end{pmatrix} \begin{pmatrix} C5 & 0 & S5 \\ 0 & 1 & 0 \\ -S5 & 0 & C5 \end{pmatrix} \begin{pmatrix} C6 & -S6 & 0 \\ S6 & C6 & 0 \\ 0 & 0 & 1 \end{pmatrix}$$

$$= \begin{pmatrix} C6C5 & -C5S6 & S5 \\ S5S4C6 + C4S6 & -S6S4S5 + C4C6 & -S4C5 \\ -C4S5C6 + S4S6 & C4S5S6 + S4C6 & C4C5 \end{pmatrix}$$

$$M_0^3 = \begin{pmatrix} C1 & -S1C2 & S1S2 \\ S1 & C1C2 & -C1S2 \\ 0 & S2 & C2 \end{pmatrix}$$

Equation (4-11) then provides the three equations:

$$S5\ell_3 = C1\,\delta_x + S1\,\delta_y$$
$$-S4C5\ell_3 = -S1C2\,\delta_x + C1C2\,\delta_y + S2\,\delta_z$$
$$C4C5\ell_3 = S1S2\,\delta_x - C1S2\,\delta_y + C2\,\delta_z \qquad \textbf{(4-12)}$$

Since θ_1, θ_2 and T have already been determined, the right-hand side consists of the constants:

$$S5\ell_3 = A$$
$$-S4C5\ell_3 = B$$
$$C4C5\ell_3 = C \qquad \textbf{(4-13)}$$

System (4-14) leads to the following solutions (θ_4^*, θ_5^*) and ($\theta_4^* + \pi$, $-\theta_5^* + \pi$) with:

$$\theta_4^* = \arctan(-B/C)$$
$$S\theta_5^* = A/\ell_3;\ C\theta_5^* = C/\ell_3 C\theta_4^*$$

The rotation through θ_6 about Z_5 cannot be determined from a knowledge of points D and E, since both are on the rotational axis of the terminal set of coordinate axes.

4. Conclusions: There are several points which should be noted:

1. Even if the number of constraints required by the operator is apparently equal to the number of DOF, their independence must be ensured. The expression $\alpha^2 + \beta^2 + \gamma^2 = 1$ implies only five constraints.
2. For an analytical solution, a method similar to the one proposed here, ie to find the end point of the arm, must be used.
3. The solutions are rarely unique, although the mechanical constraint eliminates some of them.

Now consider the example illustrated in Figure 33 which involves the manipulator shown in Figure 32 but with an end effector without intersecting axes of the type shown in Figure 31. Once again, the position of H and direction of HG are imposed in the task space Rf (which is arbitrarily identical to R_0). With these constraints, it is easy to determine the position of point D. If it is supposed that H and G in R_0 are known, the following equation can be written to describe this system:

$$\underline{OH}(R_0) = \begin{pmatrix} 0 \\ 0 \\ L_0 \end{pmatrix} + M_0^1 \begin{pmatrix} 0 \\ 0 \\ L_1 \end{pmatrix} + M_0^2 \begin{pmatrix} 0 \\ L_2 \\ 0 \end{pmatrix} + M_0^3 \begin{pmatrix} 0 \\ TL_3 \\ 0 \end{pmatrix} + M_0^4 \begin{pmatrix} 0 \\ L_4 \\ 0 \end{pmatrix} + M_0^5 \begin{pmatrix} L_5 \\ 0 \\ 0 \end{pmatrix} + M_0^6 \begin{pmatrix} 0 \\ 0 \\ -L_6 \end{pmatrix} + M_0^7 \begin{pmatrix} 0 \\ -L_7 \\ 0 \end{pmatrix} \quad (4\text{-}14)$$

$$\underline{GH}(R_0) = M_0^7 \; [\, 0, -L_7, 0 \,]^T \quad (4\text{-}15)$$

Referring to Figure 33 for the definition of the sets of coordinate axes, the following transformation matrices can be written:

$$M_0^1 = \begin{pmatrix} C1 & -S1 & 0 \\ S1 & C1 & 0 \\ 0 & 0 & 1 \end{pmatrix} \quad M_1^2 = \begin{pmatrix} 1 & 0 & 0 \\ 0 & C2 & -S2 \\ 0 & S2 & C2 \end{pmatrix} \quad M_2^3 = (\Uparrow)$$

$$M_3^4 = \begin{pmatrix} C4 & -S4 & 0 \\ S4 & C4 & 0 \\ 0 & 0 & 1 \end{pmatrix} \quad M_4^5 = (\Uparrow) \quad M_5^6 = \begin{pmatrix} 1 & 0 & 0 \\ 0 & C5 & -S5 \\ 0 & S5 & C5 \end{pmatrix}$$

$$M_6^7 = \begin{pmatrix} C6 & 0 & S6 \\ 0 & 1 & 0 \\ -S6 & 0 & C6 \end{pmatrix}$$

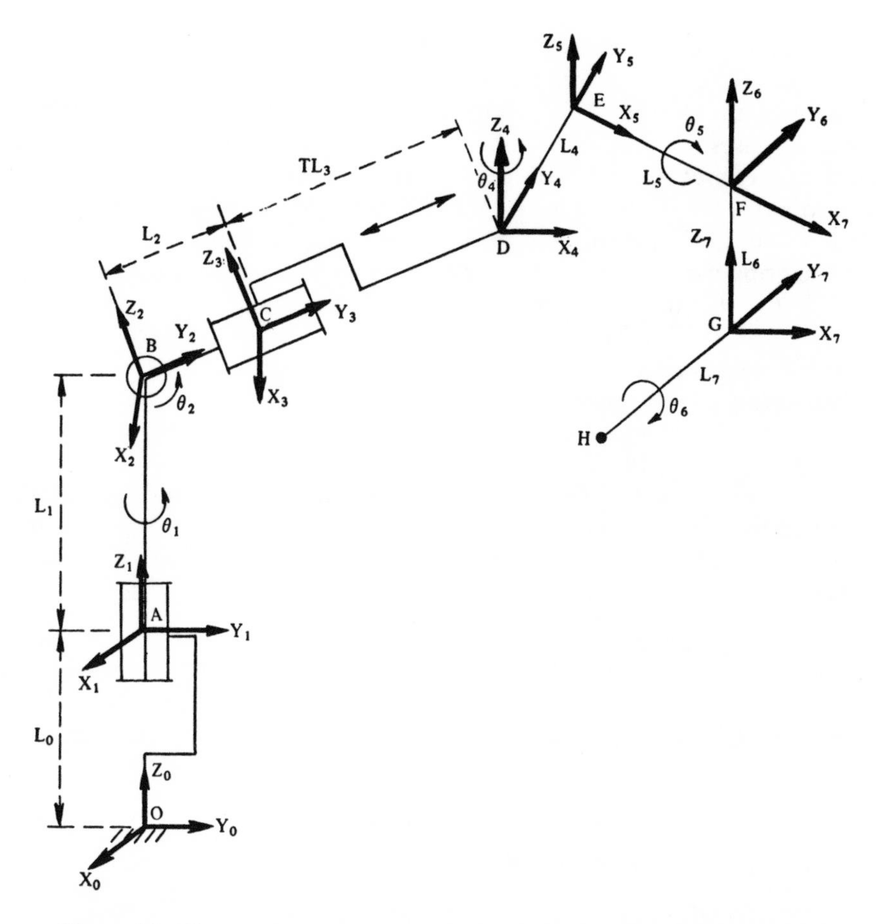

Figure 33. *The structure of an arm without intersecting end effector axes*

Equations (4-14) and (4-15) provide the following six equations (of which five are independent because H and G are on the rotational axis of θ_6):

$$X_H(R_0) = -L_2 S1C2 - TL_3 S1C2 - L_4(C1S4+S1C2C4) + L_5(C1C4 - S1C2S4) -L_6[S5(C1S4+S1C2C4) + C5S1S2] - L_7[S1S2S5-C5(S4C1+S1C2C4)] \quad \textbf{(4-16)}$$

$$Y_H(R_0) = L_2 C1C2 + TL_3 C1C2+L_4(C1C2C4-S1S4)+L_5(S1C4+C1C2S4) + L_6[S5(C1C2C4-S1S4) + C1S2C5] - L_7[C5(C1C2C4-S1S4)-C1S2S5] \quad \textbf{(4-17)}$$

$$Z_H(R_0) = L_0 + L_1 + L_2 S2 + TL_3 S2 + L_4 S2C4 + L_5 S2S4 - L_6(C2C5-S2S5C4) - L_7(S2C4C5+S5C2) \quad \textbf{(4-18)}$$

$$X_H - X_G = L_7[C5(C1S4+S1C2C4) + S5S1S2] \quad \textbf{(4-19)}$$

$$Y_H - Y_G = L_7[C1S2S5-C5(C1C2C4-S1S4)] \quad \textbf{(4-20)}$$

$$Z_H - Z_G = -L_7(S2C4C5 + S5C2) \quad \textbf{(4-21)}$$

Consider equations (4-16) to (4-20); the left-hand sides of each equation have known values – a situation which is similar to that discussed in section 4.1. Inversion of these equations (for which, in certain cases, a limited number of solutions is known) is difficult, and sometimes is impossible. A direct consequence of this is that coordinates of H and G as the set $[P_i(R_0), S_j(R_0)]$ must not be chosen. It is essential that a suitable position of D and an appropriate direction of the gripper are selected. In this case an algorithmic solution exists.

For θ_1, θ_2 and TL_3 a solution can be obtained by using equations (4-5) to (4-9). For θ_4 and θ_5 (it not being possible to determine θ_6 for reasons discussed in section 4.5, see point *3*) the following system of equations can be obtained:

$$[\alpha, \beta, \gamma]^T = M_0^7 [0, -1, 0]^T \tag{4-22}$$

This expresses the fact that the orientation of the gripper in R_0, defined by the cosines directrix α, β and γ, is equal to the orientation of the gripper in set R_7, to which the transformation, M_0^7, can be applied. Then, for example:

$$\alpha = C5\,(C1S4 + S1C2C4) - S5S1S2 \tag{4-23}$$

$$\beta = C1S2S5 - C5\,(C1C2C4 - S1S4) \tag{4-24}$$

$$\gamma = -(S2C4C5 + S5C2) \tag{4-25}$$

Since the first three articulated variables are known, this system has the form:

$$\alpha = C5\,(aS4 + bC4) - cS5 \tag{4-26}$$

$$\beta = dS5 - C5\,(eC4 - fS4) \tag{4-27}$$

$$\gamma = -(gC5 + hS5) \tag{4-28}$$

These equations have a limited number of analytical solutions which, with reference to equation (4-28), are easy to derive:

$$C5 = \pm[(h - 2\gamma g - \gamma^2)/(h^2 + g^2)]^{1/2}$$

In order to determine θ_6 the direction of X_7 or Z_7 must be known. In conclusion, in order to determine the articulated variables in a classical system possessing six DOF, both the tip of the arm and the two directions parallel to the coordinate axes of the gripper must be fixed. This corresponds to six independent constraints: the three coordinates of the point in R_0 and the three of the six cosines directrix which are independent. The following equations then follow:

$$\alpha^2 + \beta^2 + \gamma^2 = 1 \tag{4-29}$$

$$\alpha'^2 + \beta'^2 + \gamma'^2 = 1 \tag{4-30}$$

$$\alpha\alpha' + \beta\beta' + \gamma\gamma' = 0 \tag{4-31}$$

4.6 Mechanisms with more than six degrees of freedom

In Chapter 2 reasons were discussed why sometimes more than six DOF are used to produce a hierarchy of DOF involving other components, eg the vehicle, arm and end effector. Every component has, at most, six DOF and each must be considered to be separate from both the preceding one and the subsequent one.

In these cases the situations described in section 4.5 arise. However, here the constraints imposed on each subpart itself are compatible with other subparts (see Figure 34).

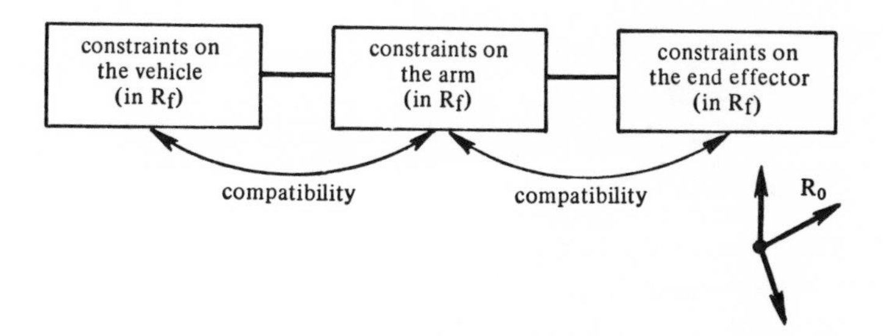

Figure 34. *The problem of compatibility for robots possessing more than six DOF*

Conclusions

The calculation of the variables in an AMS creates problems of transformation of coordinates between the task space and the articulated variables. Solution of this is essential before positional control is possible and this will be studied further in Chapter 5. It raises problems which are difficult to overcome but which must be considered during the design of a robot, in so far as it is desired to derive automatic control from a knowledge of the task space.

Chapter 5

Positional control of articulated robots

The main problem in controlling a robot can be stated thus: with the AMS in any possible configuration (which means that the articulated variables, θ_i, have any realizable value), the end effector (ie its coordinate reference axes) must be placed at a point in space with a definite orientation, relative to a fixed coordinate set (R_f) related to the environment – the coordinate set (R_0) related to the base of the robot is known relative to R_f.

In such a case every articulated variable (θ_i) must be known relative to a chosen point of origin, θ_{i0}. θ_{i0} constitutes the reference configuration for the computations and can be represented by the vector $\underline{\Theta}_0 = (\theta_{10}, \theta_{20} \ldots \theta_{n0})$, where n is the number of DOF.

The value of θ_i^* must be determined with the end effector in the required position (θ_i) – the set of coordinate axes bound to the end effector is known (see Chapter 4).

It must be possible to change each value of θ_i to equal θ_i^*; this implies the use of:

(a) the various torques or forces which act on the mechanical structure.
(b) the control system, ie the system which, given all the available information, will generate the activating signals for the actuators, and will make θ_i attain and maintain the value θ_i^*.

One important consequence of this is related to the order, range and interrelationship of the movements made and the relevant DOF. The coordinate set of axes bound to the end effector will describe a trajectory which is dependent on how the final position and orientation of the effector is obtained. Trajectory control is a problem which will be discussed in some detail in Chapter 9.

5.1 Reference and starting configurations

A simplified representation of the control of an articulation is shown in Figure 35. The principle is very simple: in the axis of the articulation there is a sensor which produces a voltage proportional to θ. This voltage opposes the control voltage (V_C) and causes the system to move towards the equilibrium position $\epsilon = 0$.

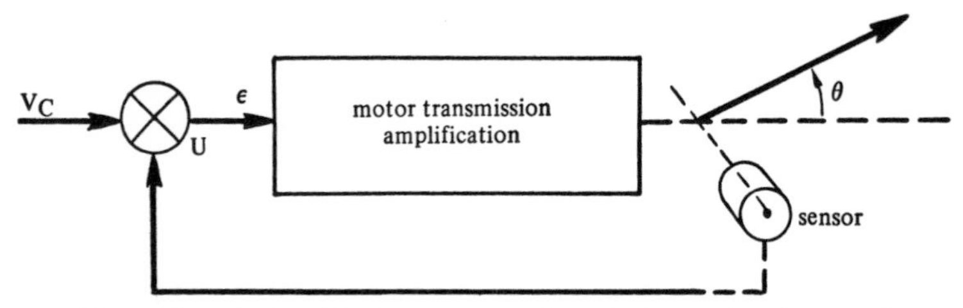

Figure 35. *A simplified representation of the mode of control of an articulation*

Assuming appropriate amplification, there is a monotonic relationship between V_C and θ. A point of origin must be chosen which will allow the evaluation of θ (the geometrical and mathematical considerations discussed in Chapter 3 should be applied here). The condition $\theta = 0$ will not necessarily correspond with $V_C = 0$. $\theta = 0$ is the *reference configuration* of the articulated variables and is not necessarily attainable (see Figure 36).

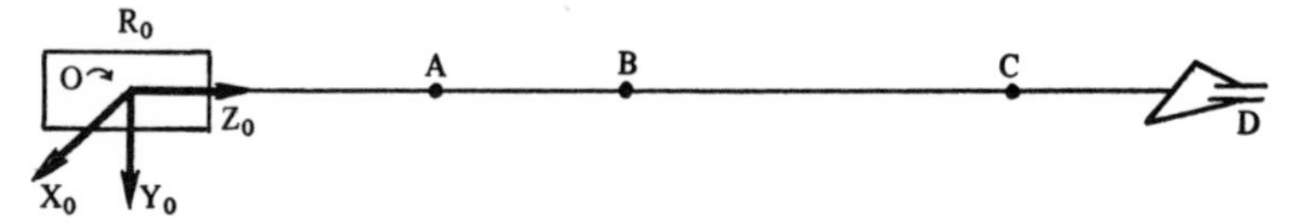

Figure 36. *An unattainable configuration of the manipulator illustrated in Figure 20*

By placing the sensors in a convenient position and adjusting them as necessary, a configuration should be obtained corresponding to $V_C = 0$. This is the *main initialization configuration* (see Figure 37) and is always attainable.

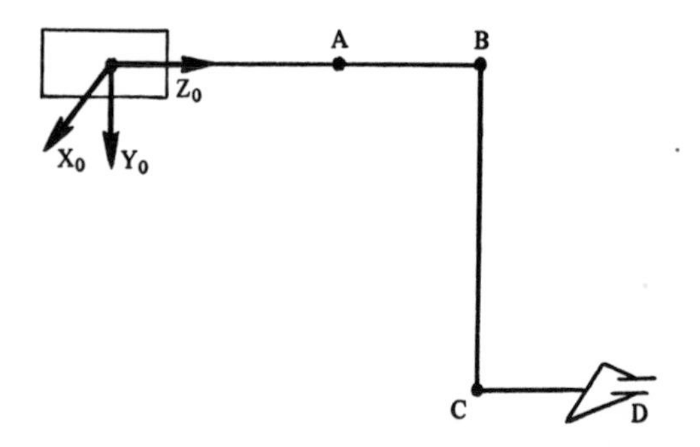

Figure 37. *The main initialization configuration of the manipulator illustrated in Figure 20:* $\underline{\Theta}_0 = (0, 0, -\pi/2, 0, \pi/2, 0)$

It may also be necessary to define a configuration in which V_C does not equal zero; such a state is called the *initialization configuration* (ie it is not the main one).

5.2 The principles of positional control

In the previous section the problem of the relationship between the control voltage and the values of the articulated variables has been solved. It is now necessary to deal with the position and orientation of the reference axes of the end effector using a vector ($\underline{X}$), bound to the end effector, relative to the set of coordinate axes R_0 (which, at most, will have six independent components).

In Chapter 3, the following monotonic relationship was described:

$$\underline{X}(R_0) = \underline{F}(\underline{\Theta} - \underline{\Theta}_0) \quad \textbf{(5-1)}$$

This represents a non-linear system relative to θ_i, $\underline{\Theta}_0$ being the main initialization configuration. Under certain conditions (see Chapter 4) equation (5-1) can be reversed to yield:

$$\underline{\Theta} - \underline{\Theta}_0 = \underline{F}^{-1}\,[\underline{X}(R_0)] \quad \textbf{(5-2)}$$

If equation (5-2) has a unique solution then it represents a positional control strategy.

Example: Consider the class 4 carrier illustrated in Figure 38.

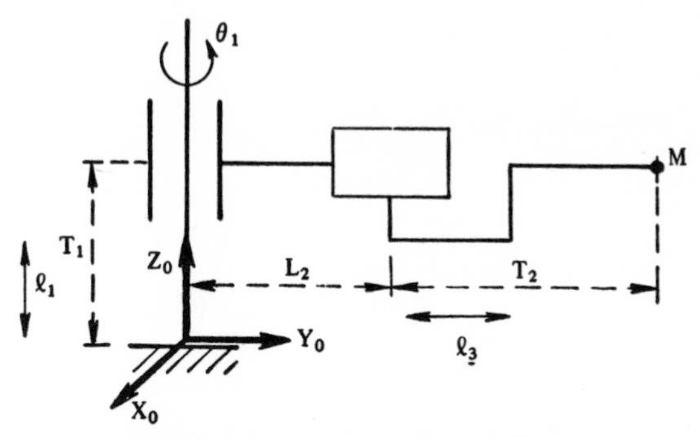

Figure 38. *A P-R-P type carrier*

Suppose that each of the three articulated variables T_1, θ_1 and T_2 is independently motor actuated (for example, electrically) by gearing down. As the main initialization configuration the vector:

$$\underline{\Theta}_0 = (\ell_1, 0, \ell_3)^T$$

can be adopted [the vector of the articulated variables being $\underline{\Theta}$ $(= T_1, \theta_1, T_2)^T$], which corresponds to the vector of the actuator variables

$\epsilon_0 = (\epsilon_{01}, \epsilon_{02}, \epsilon_{03})^T$ and also the control vector $\underline{V}_0 = (0, 0, 0)^T$ (see Figure 39):

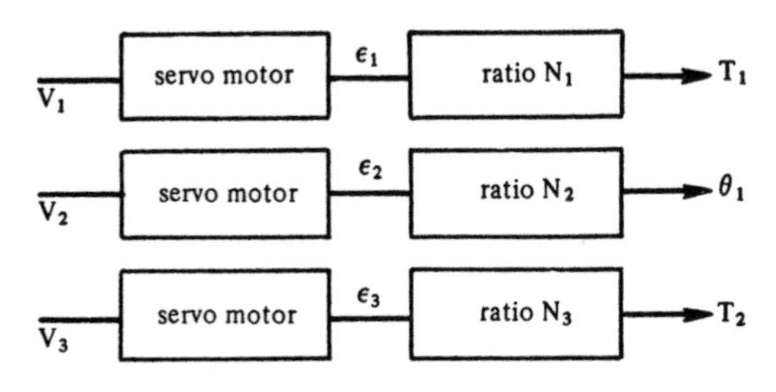

Figure 39. *The articulated variable-control voltage relationship*

The following direct relationships can be obtained:

$$K_1 V_1 = (\epsilon_1 - \epsilon_{01}) = N_1(T_1 - \ell_1)$$
$$K_2 V_2 = (\epsilon_2 - \epsilon_{02}) = N_2 \theta_1$$
$$K_3 V_3 = (\epsilon_3 - \epsilon_{03}) = N_3(T_2 - \ell_3)$$

Taking the known coordinates of M in R_0 for the vector $\underline{X}$:

$$\underline{X}(R_0) = (X_M^0, Y_M^0, Z_M^0)^T$$

The geometry of the system allows the following expressions to be written:

$$X_M^0 = -(L_2 + T_2)\,C1$$
$$Y_M^0 = (L_2 + T_2)\,S1$$
$$Z_M^0 = T_1 \quad \textbf{(5-3)}$$

Inversion of equation (5-3) is easy to perform and, assuming that $0 < \theta_1 < \pi$, provides:

$$T_1 = Z_M^0$$
$$\theta_1 = \arctan(-Y_M^0/X_M^0)$$
$$T_2 = -L_2 + Y_M^0/S1 \quad \textbf{(5-4)}$$

from which:

$$\underline{\Theta}^* - \underline{\Theta}_0 = [Z_M^0 - \ell_1, -\arctan(Y_M^0/X_M^0), -L_2 - \ell_3 + (Y_M^0/S\theta_1)]^T \quad \textbf{(5-5)}$$

$$(V_1^*, V_2^*, V_3^*) = [\frac{N_1}{K_1}(Z_M^0 - \ell_1), -\frac{N_2}{K_2}\arctan\left(\frac{Y_M^0}{X_M^0}\right), -\frac{N_3}{K_3}(L_2 + \ell_3 - \frac{Y_M^0}{S\theta_1})] \quad \textbf{(5-6)}$$

By applying the control vector (V_1^*, V_2^*, V_3^*) it is possible to go from configuration $\underline{\Theta}_0$ to configuration $\underline{\Theta}$, which satisfies the constraints of $\underline{X}(R_0)$.

5.3 Balanced and initialization configurations

When the actuators are no longer energized, the articulated system will assume a configuration which is imposed by gravity, the nature of the actuators and the transmissions to the various joints. For irreversible transmissions and actuators the configuration is balanced when the power supply to the actuators is switched off. For a reversible system, it is gravity which determines the balanced configuration(s).

To avoid collapse of a de-energized system, a mechanical system of compensation (see Figures 40 and 41) can be used, which can take, for instance, the shape of a flexible parallelogram. The gravitational torques arising from the segments involved in the first three DOF are compensated for in such a way that, neglecting the mass of the end effector (the effect of which is partly compensated for by dry friction), an undifferentiated balance for a commonly used configuration is obtained.

Example: Suppose that in the system illustrated in Figure 40[14] OABC are in the vertical plane and BB′VV′ is a flexible parallelogram. The body AB has a mass M_2 and is represented by its mass centre G_2. BC has a mass M_3 and G_3 is its mass centre. The counterweight of mass M has its mass centre in S; the bodies VV′ and V′B have negligible masses, compared with those in AB and BC.
In the set of coordinate axes YAZ, AY representing the vertical, the Y coordinates of the mass centres are:

$$Y_S = VS.S(2+3) - AV.S2 \quad (5\text{-}7)$$

$$Y_{G_2} = 1/2.AB.S2 \quad (5\text{-}8)$$

$$Y_{G_3} = AB.S2 - 1/2.BC.S(2+3) \quad (5\text{-}9)$$

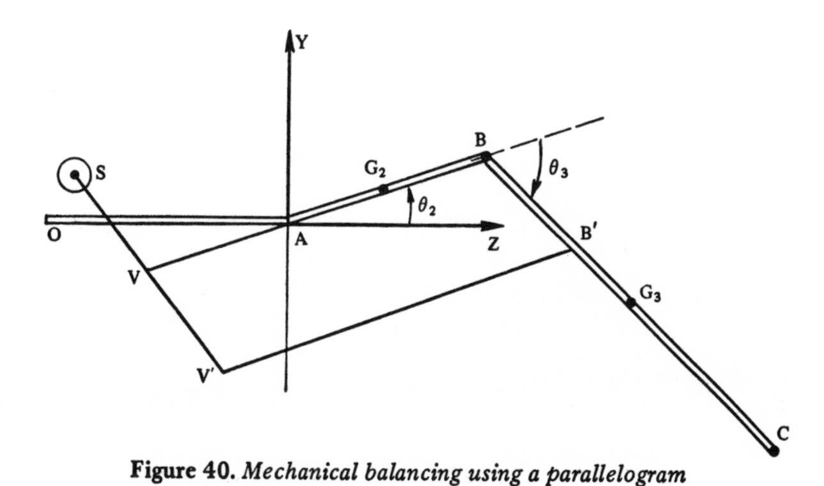

Figure 40. *Mechanical balancing using a parallelogram*

The gravitational torques arising in A and B are:

$$Q_A = g\left(M\frac{\partial Y_S}{\partial \theta_2} + M_2\frac{\partial Y_{G_2}}{\partial \theta_2} + M_3\frac{\partial Y_{G_3}}{\partial \theta_2}\right) \quad (5\text{-}10)$$

$$Q_B = g\left(M\frac{\partial Y_S}{\partial \theta_3} + M_2\frac{\partial Y_{G_2}}{\partial \theta_3} + M_3\frac{\partial Y_{G_3}}{\partial \theta_3}\right) \quad (5\text{-}11)$$

Q_A and Q_B cancel each other out if:

$$M.VS = M_3 . BC/2 \quad (5\text{-}12)$$
$$M.AV = M_2 . AB/2 + M_3 . AB \quad (5\text{-}13)$$

Equations (5-12) and (5-13) allow, by imposing a value of M, the determination of VS and AV.

If the mechanical system illustrated in Figure 40 is rotated through θ_1 about OA, the system shown in Figure 41 is obtained. It should be noted that the mechanical arrangement forces the use of two parallelograms BB′VV′ placed symmetrically relative to the plane YAZ.

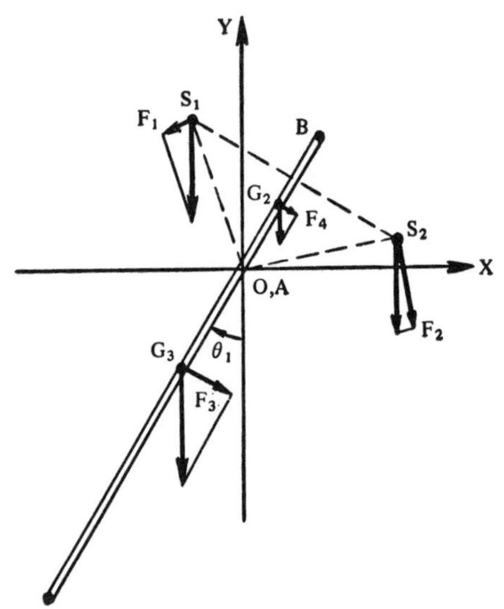

Figure 41. *Mechanical balancing of an azimuthal rotation*

Examination of this figure shows that the torques resulting from S_1 and S_2, and which act on A, can be seen to have an effect in the opposite direction, additional to the torques resulting from G_2 and G_3. The difference can be small, even for a wide range of values of θ_1 (and of θ_2 and θ_3), and might, in fact, be compensated for by friction.

It should be noted that F_2 is greater than F_1, which will help compensate for the difference between the torques arising from F_2 and F_3. Adjustment of the distance S_1 to S_2 allows precise compensation for one configuration, and which may be sufficient for many other relatively similar configurations.

Whatever the AMS, the de-energized balance configuration can, in certain conditions, be either the reference configuration or the main initialization configuration.

This has a direct bearing on the practical problem concerning the start-up of a robot. As explained in section 5.1, the actuators (being permanently servoed) are, in fact, controlled by the difference between the signal from the positional sensors and the control signal (see Figure 35). Suppose the system to be at rest (actuators off) and that the sensors and the actuators are fed simultaneously. What then happens? The control signal (V_C) being zero, the true control signal is $\epsilon = V_C = U$. Unless the system is in the main initialization configuration (which is unlikely), U is very different from V_C. The servoing will tend to cancel out ϵ, and therefore, U. The robot will move towards its main initialization configuration at maximum speed – apart from putting the system under stress, this is a safety risk (and also the robot could be damaged).

Several methods can be used to overcome this difficulty. One of the easiest to apply is to first supply the sensors and then measure U and apply a voltage (V_C) which is equal to U and reduces ϵ to zero. Then the actuators can be supplied; in principle, the robot remains in its original configuration which becomes the initialization configuration for the task to be performed.

This is very useful in practice since any balanced configuration can become an initialization configuration.

5.4 The problems associated with positional control

In this type of control, the main initialization configuration ($\underline{\Theta}_0$), the starting (or initialization) configuration ($\underline{\Theta}_{dp}$) and the required configuration ($\underline{\Theta}_{df}$) are known. The relation between the control vector ($\underline{V}$) and the configuration $\underline{\Theta}$ resulting from the displacement is also known.

Suppose the system is at rest in configuration $\underline{\Theta}_{dp}$ and is subjected to the control vector $\underline{V}_{dp}$. In order that the robot may take up configuration $\underline{\Theta}_{df}$ a control step vector of amplitude $\underline{\Delta V}_{pf} = \underline{V}_{df} - \underline{V}_{dp}$ must be applied. The response of the mechanical system to this vector (supposing that the individual actuators are energized simultaneously) is to obey the dynamic equations of the system. As will be shown in Chapter 7, these equations are complex and non-linear. If the amplitude demand of the variation $\underline{\Theta}$ is too great, oscillations might occur which may cause the robot to achieve an inaccurate final configuration (a typical response of a non-linear system to a step input). The speed of movement of the mechanical system must be limited, and/or small, successive increments (the amplitude of which can increase, for example, according to the distance of the instantaneous configuration from the final configuration) be applied to the control.

These points will now be illustrated by a simple but important example (see Figure 42).

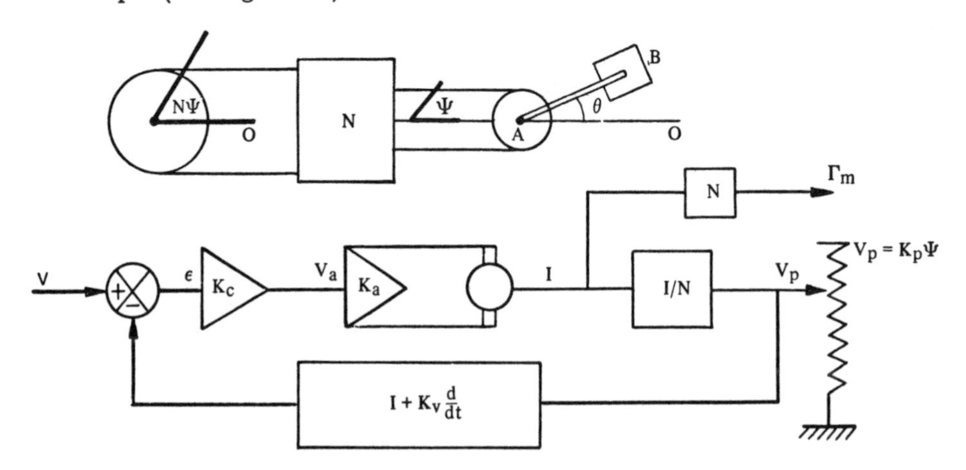

Figure 42. *A system possessing one DOF and with servo control*

An electrical torque-motor activates a limb (AB) through a reduction gear without backlash, friction or elasticity and with negligible inertia. The motor is servoed by a potentiometer connected to the output of the reduction gear. The control signal (ϵ) is amplified to energize the motor, drawing a current (I), and theoretically giving a torque proportional to I. This motor acts across the reduction gear on the sensor potentiometer, and the filtered signal (proportional and derivative filter) will counteract the input voltage (V).

If Ψ is the motor variable (with $\Psi_0 = \theta_0 = 0$), N the step-down ratio of the gear and θ, the articulated variable, then:

$$V_p = K_p \Psi \tag{5-14}$$

$$V_a = K_c [V - (1 + K_v \frac{d}{dt})] \Psi \tag{5-15}$$

$$I = K_a V_a \tag{5-16}$$

$$\Gamma_m = N K_T I \tag{5-17}$$

where K_p is the potentiometer constant and K_T the torque constant.

The theoretical torque (Γ_m) will then be:

$$\Gamma_m = K_T K_a N [K_c V - K_c K_p \Psi - K_c K_v K_p \frac{d\Psi}{dt}] \tag{5-18}$$

If the motor has an inertia j and a viscous friction with coefficient f, the theoretical torque is affected and the useful torque acting on the lever arm becomes:

$$\Gamma_\Psi = \Gamma_m - j N^2 \frac{d^2 \Psi}{dt^2} - f N^2 \frac{d\Psi}{dt} \tag{5-19}$$

Suppose:

$$\Gamma_{\Psi} = K_T\, K_a\, K_c\, NV - K_T\, K_a\, K_c\, K_p N\, \Psi - (K_T\, K_a\, K_c\, K_v\, K_p\, N + f\, N^2)\frac{d\Psi}{dt} - j\, N^2\,\frac{d^2\Psi}{dt^2} \tag{5-20}$$

but, on the other hand, this calculation can be simplified to:

$$\Gamma_{\Psi} = \alpha\, V - \beta\, \Psi - \gamma\, \frac{d\,\Psi}{dt} - \delta\, \frac{d^2\Psi}{dt^2} \tag{5-21}$$

Suppose that $\Psi = \theta$:

$$\Gamma_{\theta} = \alpha\, V - \beta\, \theta - \gamma\, \frac{d\theta}{dt} - \delta\, \frac{d^2\theta}{dt^2} \tag{5-22}$$

It is very easy to apply the Lagrange equation to lever MB (see Chapter 7):

$$j\, \frac{d^2\theta}{dt^2} + R\left(\frac{d\theta}{dt}\right)^2 = Q_{\theta} + \Gamma_{\theta} \tag{5-23}$$

where j is the inertia of the lever with respect to its axis of rotation, R is a constant which is associated with the centrifugal force, Γ_θ is the active torque, and Q_θ the gravitational torque. To evaluate this torque it can be supposed that the mass (M) is concentrated in B; then:

$$Q_{\theta} = AB.\, Mg \cos\theta \tag{5-24}$$

By combining equations (5-22) to (5-24) the evolution equation of θ can be obtained:

$$(j + \delta)\, \frac{d^2\theta}{dt^2} + R\left(\frac{d\theta}{dt}\right)^2 + \gamma\, \frac{d\theta}{dt} + \beta\, \theta - AB.Mg.\cos\theta = \alpha\, V \tag{5-25}$$

In positional control a hypothesis is made that a progressive succession of states of equilibrium occurs for which:

$$(d\theta/dt) = (d^2\theta/dt^2) = 0$$

The control equation obtained is:

$$V = (\beta\theta - AB.\, Mg \cos\theta)/\alpha$$

Since the required angle is known, the value of V which needs to be applied to achieve the final angle is also known. Equation (5-25) gives an indication of the risks involved in obtaining positional control; these risks increase when the change in θ is great.

Of the methods used to overcome these difficulties, particular note should be made of the method which makes use of transfer functions (or filters) (or non-integer order and < 2) in the servo-loops of the motor. These filters can be made using properly matched resistance-capacitor networks.[15] Then the transitory response of the servos remains reasonably constant, despite even considerable variations in the inertial motor load (up to 300 per cent of the nominal inertia according to Oustaloup). This method can substantially improve the limits of positional control.

Chapter 6

Speed control of articulated robots

Returning to:

$$\underline{X}(R_0) = \underline{F}(\underline{\Theta}-\underline{\Theta}_0)$$

it should be noted that knowing the values of the articulated variables ($\underline{\Theta}$) in a configuration with an origin of $\underline{\Theta}_0$, a monotonic relationship exists between $\underline{\Theta}$, $\underline{\Theta}_0$ and the parameters. This makes it possible to define the coordinate axes of the end effector [the parameters being gathered in a vector ($\underline{X}$) of which six components are independent] in a coordinate reference set R_0 (related to the stationary base of the robot) and the values of $\underline{\Theta} - \underline{\Theta}_0$.

In Chapter 4 the difficulties involved in inverting equation (5-1) were discussed and the problems caused by the direct use of this formula in positional control have been discussed in Chapter 5. Here, another type of control (*speed control*) will be studied, which is based on incremental variations of the values of $\underline{X}$ and $\underline{\Theta}$.

6.1 The principles of speed control

Differentiating equation (5-1):

$$\underset{(n \times 1)}{\underline{\Delta X}(R_0)} = \underset{(n \times m)}{\left(\frac{\partial \underline{F}}{\partial \underline{\Theta}}\right)} \cdot \underset{(m \times 1)}{\underline{\Delta \Theta}} \qquad \textbf{(6-1)}$$

$$\left[\frac{\partial \underline{F}}{\partial \underline{\Theta}}\right] = [J(\underline{\Theta})] \qquad \textbf{(6-2)}$$

Equation (6-2) is the Jacobian matrix of function $\underline{F}$. If equation (5-1) is developed further the following expressions can be obtained:

$$\begin{aligned} X_1 &= f_1(\theta_1 \cdots \theta_m) \\ X_2 &= f_2(\theta_1 \cdots \theta_m) \\ &\vdots \\ X_n &= f_n(\theta_1 \cdots \theta_m) \end{aligned} \qquad \textbf{(6-3)}$$

$$\begin{pmatrix} \Delta X_1 \\ \Delta X_2 \\ \vdots \\ \Delta X_n \end{pmatrix} = \begin{pmatrix} \frac{\partial f_1}{\partial \theta_1} & \frac{\partial f_1}{\partial \theta_2} & \cdots & \frac{\partial f_1}{\partial \theta_m} \\ \frac{\partial f_2}{\partial \theta_1} & \frac{\partial f_2}{\partial \theta_2} & \cdots & \frac{\partial f_2}{\partial \theta_m} \\ \frac{\partial f_n}{\partial \theta_1} & \frac{\partial f_n}{\partial \theta_2} & \cdots & \frac{\partial f_n}{\partial \theta_m} \end{pmatrix} \begin{pmatrix} \Delta \theta_1 \\ \Delta \theta_2 \\ \vdots \\ \Delta \theta_m \end{pmatrix} \quad (6\text{-}4)$$

$$n \times 1 \qquad n \times m \qquad m \times 1$$

of which a condensed expression is:

$$\underset{n \times 1}{\underline{\Delta X}} = \underset{n \times m}{[J]} \; \underset{m \times 1}{\underline{\Delta \Theta}} \quad (6\text{-}5)$$

Reference to set R_0 will be implicit from now on.

Example: Referring to the arm shown in Figure 28, to which equation (4-1) was applied:

$$X_C = -S1C2\,(L_2 + T) \quad (6\text{-}6)$$

$$Y_C = C1C2\,(L_2 + T) \quad (6\text{-}7)$$

$$Z_C = L_1 + S2\,(L_2 + T) \quad (6\text{-}8)$$

By differentiating these relations and using equation (6-4), the following can be obtained:

$$\Delta X_C = -C1C2(L_2+T).\Delta\theta_1 + S1S2(L_2+T).\Delta\theta_2 - S1C2.\Delta T \quad (6\text{-}9)$$

$$\Delta Y_C = -S1C2(L_2+T).\Delta\theta_1 - C1S2(L_2+T).\Delta\theta_2 + C1C2.\Delta T \quad (6\text{-}10)$$

$$\Delta Z_C = C2(L_2+T).\Delta\theta_2 + S2.\Delta T \quad (6\text{-}11)$$

A Jacobian matrix can then be written:

$$J = \begin{pmatrix} -C1C2\,(L_2+T) & S1S2\,(L_2+T) & -S1C2 \\ -S1C2\,(L_2+T) & -C1S2\,(L_2+T) & C1C2 \\ 0 & C2(L_2+T) & S2 \end{pmatrix} \quad (6\text{-}12)$$

which can be inverted to allow calculation of the increments $\Delta\theta_1$, $\Delta\theta_2$ and ΔT:

$$C2(L_2+T)\Delta\theta_1 = -C1\Delta X_C - S1\Delta Y_C \quad (6\text{-}13)$$

$$(L_2+T)\Delta\theta_2 = S1S2\Delta X_C - C1S2\Delta Y_C + C2\Delta Z_C \quad (6\text{-}14)$$

$$\Delta T = -S1C2\Delta X_C + C1C2\Delta Y_C + S2\Delta Z_C \quad (6\text{-}15)$$

If the initial values of X_C, Y_C, Z_C, θ_1, θ_2 and T, and the desired increments ΔX_C, ΔY_C and ΔZ_C are known, then equations (6-10) to (6-12) can be used to determine the values of the increments of the articulated variables $\Delta\theta_1$, $\Delta\theta_2$ and ΔT. A general equation of speed control can then be written as:

$$\underset{m \times 1}{\underline{\Delta \Theta}} = \underset{m \times n}{[J]^{-1}} \; \underset{n \times 1}{\underline{\Delta X}} \quad (6\text{-}16)$$

6.2 Problems arising from the use of equation (6-16)

Matrix [J] is classically reversible only if:

(a) it is a square array with equal m and n dimensions of vectors $\underline{\Delta\Theta}$ and $\underline{\Delta X}$.
(b) its determinant is not zero; this is dependent only on the value of the articulated variables, ie of the configuration of the robot. (The rank of J is m = n.)

These two conditions are not satisfied under the following circumstances:

1. J is not square because n < m might apply: Consider a movement where ΔX_C and ΔZ_C are imposed on the system shown in Figure 28. Then equations (6-9) to (6-11) can be reduced to:

$$\underset{2 \times 1}{\begin{pmatrix} \Delta X_C \\ \Delta Z_C \end{pmatrix}} = \underset{2 \times 3}{\begin{pmatrix} -C1C2(L_2+T) & S1S2(L_2+T) & -S1C2 \\ 0 & C2\,(L_2+T) & S2 \end{pmatrix}} \underset{3 \times 1}{\begin{pmatrix} \Delta\theta_1 \\ \Delta\theta_2 \\ \Delta T \end{pmatrix}} \qquad (6\text{-}17)$$

If ΔX_C, ΔZ_C, θ_1, θ_2 and T are known and the rank of matrix J is 2, then the triplet $\Delta\theta_1$, $\Delta\theta_2$ and ΔT has an infinite number of solutions which satisfy equation (6-17). To reverse this equation and to fix one of the variables entails the use of redundant DOF.

2. *The constraints imposed on vector $\underline{\Delta X}$ are independent and the determinant of J is zero:* Suppose that in equations (6-9) to (6-11) the starting configuration is $\theta_2 = \pi/2$. The system can then be expressed as:

$$\begin{pmatrix} \Delta X_C \\ \Delta Y_C \\ \Delta Z_C \end{pmatrix} = \begin{pmatrix} 0 & S1(L_2+T) & 0 \\ 0 & -C1(L_2+T) & 0 \\ 0 & 0 & 1 \end{pmatrix} \begin{pmatrix} \Delta\theta_1 \\ \Delta\theta_2 \\ \Delta T \end{pmatrix} \qquad (6\text{-}18)$$

which is equal to:

$$\begin{pmatrix} \Delta X_C \\ \Delta Y_C \\ \Delta Z_C \end{pmatrix} = \begin{pmatrix} S1(L_2+T) & 0 \\ -C1(L_2+T) & 0 \\ 0 & 1 \end{pmatrix} \begin{pmatrix} \Delta\theta_2 \\ \Delta T \end{pmatrix} \qquad (6\text{-}19)$$

This equation can be solved only if ΔX_C and ΔY_C are related through:

$$\Delta X_C = -\Delta Y_C . \tan\theta_1 \qquad (6\text{-}20)$$

Unless this relation is satisfied (which is generally difficult since ΔX_C, ΔY_C and ΔZ_C are independent and are imposed *a priori*) equation (6-19) cannot be solved and the constraint becomes incompatible with the configuration.

3. The determinant of J is 0 because the constraints on $\underline{\Delta X}$ are not independent: If equation (6-20) holds true, then equation (6-18) can be written as:

$$\begin{pmatrix} \Delta Y_C \\ \Delta Z_C \end{pmatrix} = \begin{pmatrix} -C1(L_2 + T) & 0 \\ 0 & 1 \end{pmatrix} \begin{pmatrix} \Delta\theta_2 \\ \Delta T \end{pmatrix} \quad (6\text{-}21)$$

$$\Delta X_C = \Delta Y_C . \tan\theta_1 \quad (6\text{-}22)$$

An infinite number of solutions then exists which are defined by:

$$\Delta\theta_2 = -\Delta Y_C / C1(L_2+T) = \Delta X_C / S_1(L_2+T)$$

$$\Delta T = \Delta Z_C$$

$\Delta\theta_1$ is indeterminate and arbitrary

The configuration (θ_1, $\theta_2 = \pi/2$, T) is termed *singular*. According to the constraints imposed on the movement, a situation occurs where there is either an absence of a solution (see *2* above), or an infinite number of solutions (see *3* above).

When the constraints are compatible, the problem which must be solved in practice is that of redundancy, ie the number of independent constraints is less than the number of independent DOF. Such a situation is equivalent to $n < m$ or to the rank $J \leqslant n < m$.

6.3 Methods of resolving redundant systems[16]

Recalling the method of indeterminate coefficients,[17] equation (6-5) can be converted into a homogeneous form:

$$\underset{m' \times n'}{[Q]} \quad \underset{n' \times 1'}{\underline{\Delta Z}} = \underline{0}$$

If the m′ equations are independent there will be $C_n^{m'+1}$ arbitrary coefficients which, eventually, will allow the criterion to be optimized. The drawback to this method lies in the constraints which define the independent movements. Also, the calculation is lengthy because all the determinants of rank m′ in J must be evaluated.

6.3.1 PRINCIPAL VARIABLE ANALYSIS[18]

This is a very attractive method because it is simple. It consists of leaving some variables fixed, allowing matrix J to be square. Returning to the system described by equation (6-17), having two independent constraints and three variables, and taking θ_1 and T as the main variables (θ_2 = constant; $\Delta\theta_2 = 0$):

$$J_{\theta_2} = \begin{pmatrix} -C1C2(L_2+T) & -S1C2 \\ 0 & S2 \end{pmatrix} \tag{6-23}$$

and, leaving θ_1 constant:

$$J_{\theta_1} = \begin{pmatrix} S1S2(L_2+T) & -S1C2 \\ C2(L_2+T) & S2 \end{pmatrix} \tag{6-24}$$

If θ_1 and θ_2 are chosen as the principal variables:

$$J_T = \begin{pmatrix} -C1C2(L_2+T) & S1S2(L_2+T) \\ 0 & C2(L_2+T) \end{pmatrix} \tag{6-25}$$

J_{θ_1}, J_{θ_2}, J_T are truly square matrices. If the system is described by n equations having m unknowns, then m − n unknowns are given their previous values. Then:

$$\underset{n \times 1}{\underline{\Delta X}} = \underset{n \times n}{[J^*]} \quad \underset{n \times 1}{\underline{\Delta \Theta}}$$

As stated, C_m matrices [J*] are truly square. However, the choice of the remaining variables is limited by the following restrictions.

1. The determinant of [J*] cannot be zero for these particular values of the variables. This demonstrates that the articulated system is in a singular configuration. The rank of J* is less than n. For example, the determinant of $J_{\theta_2} = 0$ if $\theta_2 = 0$. In this case $[J_{\theta_2}]$ cannot be inverted:

   $$\Delta X_C = -C1(L_2+T)\,\Delta\theta_1 - S1\Delta T$$
   $$\Delta Z_C = 0$$

   As mentioned in section 6.2, this might suggest:

   (a) a demand for an increase in the value of Z_C which cannot be satisfied by choosing J_{θ_2}.
   (b) if the system has a rank of 1, an increase in X_C will be satisfied by an infinite number of combinations of $\Delta\theta_1$, ΔT; the same processes can then be applied to θ_1 (or T) as were applied to θ_2 – for example, leaving θ_1 constant:

   $$\Delta T = -\Delta X_C / S1$$

2. The mechanical considerations should also be taken into account when choosing the principal variables. If an articulated variable is at its limit then it is better to treat it as a principal variable only if its movement removes it from this limit.

6.3.2 CASES WHERE CHOICES ARE BOTH MECHANICALLY AND MATHEMATICALLY POSSIBLE

When the rank of J is less than m (the number of articulated variables) and when all the solutions are possible (in the previous example there was no indication as to how the choice between J_{θ_1}, J_{θ_2} and J_T was made), a solution can be sought which minimizes the number of criteria.

6.3.2.1 Quadratic criteria

These have received much attention since the solutions that minimize them can be determined in the presence of linear constraints, similar to equation (6-5). Such a quadratic criterion can be written as:

$$C = \frac{1}{2} \underline{\Delta\Theta}^T [M] \underline{\Delta\Theta} \tag{6-26}$$

[M] being a symmetrical, definite and positive matrix with the linear constraint:

$$\underline{\Delta X} = J \underline{\Delta\Theta} \tag{6-5}$$

the Lagrangian of the system is defined using:

$$L = \frac{1}{2} \underline{\Delta\Theta}^T [M] \underline{\Delta\Theta} + \underline{\Lambda}^T (\underline{\Delta X} - J \underline{\Delta\Theta}) \tag{6-27}$$

where $\underline{\Lambda}^T$ represents the vector of the Lagrange multipliers. It can be shown that[19] the minimum value of L is obtained for:

$$\underline{\Delta\Theta} = M^{-1} J^T (JM^{-1} J^T)^{-1} \underline{\Delta X} \tag{6-28}$$

Example: Let M be a unit matrix. This criterion represents the minimization of the Euclidian norm of the articulated variations.

$$C = \frac{1}{2} \underline{\Delta\Theta}^T . \underline{\Delta\Theta} = \frac{1}{2} \sum_{i=1}^{m} \Delta\theta_i^2 \tag{6-29}$$

The use of this criterion allows the displacement of the articulations to be minimized. However, as the leverage of every articulated variable increases and the distance from the base of the robot increases, this criterion becomes less ideal. Applying this criterion to equation (6-17):

$$\begin{pmatrix} \Delta X_C \\ \Delta Z_C \end{pmatrix} = \begin{pmatrix} -C1C2(L_2+T) & S1S2(L_2+T) & -S1C2 \\ 0 & C2\,(L_2+T) & S2 \end{pmatrix} \begin{pmatrix} \Delta\theta_1 \\ \Delta\theta_2 \\ \Delta T \end{pmatrix}$$

Equation (6-28) can then be reduced to:

$$\underline{\Delta\Theta} = J^T (J\,J^T)^{-1} \underline{\Delta X}. \tag{6-30}$$

or:

$$\begin{pmatrix}\Delta\theta_1\\ \Delta\theta_2\\ \Delta T\end{pmatrix}=\begin{pmatrix}-C1C2(L_2+T) & 0\\ S1S2(L_2+T) & C2(L_2+T)\\ S1C2 & S2\end{pmatrix}\left[\begin{pmatrix}C1C2(L_2+T) & S1S2(L_2+T) & -S1C2\\ 0 & C2(L_2+T) & S2\end{pmatrix}\begin{pmatrix}-C1C2(L_2+T) & 0\\ S1S2(L_2+T) & -C2(L_2+T)\\ -S1C2 & S2\end{pmatrix}\right]\begin{pmatrix}\Delta X_c\\ \Delta Z_c\end{pmatrix} \quad (6\text{-}31)$$

which, after computation, leads to the solution:

$$\delta.\Delta\theta_1 = -C1C2L(S2^2+C2^2L^2)\Delta X_C+S1S2C1C2^2(L^2-1)\Delta Z_C \quad (6\text{-}32)$$

$$\delta.\Delta\theta_2 = [S1S2L(S2^2+C2^2L^2)-S2^2S1C2(L^2-1)]\Delta X_C$$
$$+[-S1^2S2^2C2L(L^2-1)+S2(C1^2C2^2L^2+S1^2S2^2L^2+S1^2C2^2)]\Delta Z_C \quad (6\text{-}33)$$

$$\delta.\Delta T = -[S1C2(S2^2+C2^2L^2)+S1S2C2^2L(L^2-1)]\Delta X_C+[S1^2$$
$$-S2C2^2(L^2-1)+C2L(C1^2C2^2L^2+S1^2S2^2L^2+S1^2C2^2)]\Delta Z_C \quad (6\text{-}34)$$

with $L = L_2 + T$:

$$\delta = L^2[C2^2(C1^2C2^2L^2+S2^2) + S1^2(S2^2+C2^4)] \quad (6\text{-}35)$$

It should be noted that, even for this relatively simple case, the full solution is rather complex.

If the problem was described by an equation and three unknowns, the method would remain valid (the criterion is not merely an added constraint assigned to the system).

Suppose that:

$$\Delta X_C = -C1C2(L_2+T)\Delta\theta_1+S1S2(L_2+T)\Delta\theta_2-S1C2\Delta T \quad (6\text{-}36)$$

and $[M] = [\Uparrow]$

$$[J] = (-C1C2L \quad S1S2L \quad -S1C2) \quad (6\text{-}37)$$

$$JJ^T = C1^2C2^2L^2+S1^2S2^2L^2+S1^2C2^2 = A \quad (6\text{-}38)$$

$$A\Delta\theta_1 = -C1C2L\,\Delta X_C \quad (6\text{-}39)$$

$$A\Delta\theta_2 = S1S2L\,\Delta X_C \quad (6\text{-}40)$$

$$A\Delta T = -S1C2\,\Delta X_C \quad (6\text{-}41)$$

Example: Suppose that M now represents the kinetic energy matrix. In Chapter 8 a matrix, of which the elements are complex functions of the articulated variables, will be calculated. It can be assumed that this matrix has the form:

$$M = \begin{pmatrix}t_0 & 0 & t_1\\ 0 & t_2 & 0\\ t_3 & 0 & t_0\end{pmatrix}$$

and that the constraint is given by equation (6-36). Then:

$$(JM^{-1}J^T)^{-1} = D/B$$

with:

$$D = t_0^2\,t_2 - t_1\,t_2\,t_3$$

$$B = C1C2^2L(C1Lt_0t_2-S1t_2t_3)+S1^2S2^2L^2(t_0^2-t_1t_3)-S1C2^2(C1Lt_1t_2-S1t_0t_2)$$

The following, unique solution can then be obtained:

$$B\,\Delta\theta_1 = [S1t_1 - C1t_0]\,C2t_2 . \Delta X_C \quad (6\text{-}42)$$

$$B\,\Delta\theta_2 = (t_0^2 - t_1 t_3)\,S1S2L\,\Delta X_C \quad (6\text{-}43)$$

$$B\,\Delta T = [C1Lt_3 - S1t_0]C2t_2\,\Delta X_C \quad (6\text{-}44)$$

In practice, where $\Delta\theta_i$ corresponds, by definition, to the increments which make it possible to linearize the problems of control (which means that these increments are as small as possible); the minimization of quadratic criteria, such as kinetic energy, is of limited interest. This criterion is dynamic and the model chosen for control is kinematic. Use of the quadratic criterion leads to long and complex computations that will create problems in real-time control.

6.3.2.2 Non-quadratic criteria

Such criteria can be easier to use and are more significant. This is true of the weighted sum of the amplitudes of the articulated variables,[20] the criterion being related (see section 6.3.2.1) to the leverage exerted by each articulated variable:

$$C = \sum_{i=1}^{m} a_i\,|\Delta\theta_i| \quad a_i > 0 \quad (6\text{-}45)$$

To find a solution, it is necessary to arbitrarily eliminate m − (n + 1) movements when J has a dimension n x m.

Returning to the problem described by equation (6-36):

$$\Delta X_C = -C1C2(L_2+T)\Delta\theta_1 + S1S2(L_2+T)\Delta\theta_2 - S1C2\,\Delta T \quad (6\text{-}36)$$

and supposing the criterion:

$$C = L_1|\Delta\theta_1| + (L_2+T)\,|\Delta\theta_2| + |\Delta T| \quad (6\text{-}46)$$

J has the dimension 1 x 3; m − (n + 1) = 3 − (1 + 1) (= one movement) can then be arbitrarily eliminated.

For $\Delta\theta_1$ the following equation can be written:

$$\Delta\theta_2 = [C2\Delta T/S2(L_2+T)] + \Delta X_C/S1S2(L_2+T) \quad (6\text{-}47)$$

$$[(L_2+T)|\Delta\theta_2| + |\Delta T|]_{minimum} = C_{min} \quad (6\text{-}48)$$

$$C_{min} = |\cot 2\,\Delta T + (\Delta X_C/S1S2)| + |\Delta T| \quad (6\text{-}49)$$

C_{min} is represented graphically in Figure 43 as a function of ΔT. Equation (6-49) can be partitioned into two parts which can then be added together:

$$C_1 = |\Delta T|$$

$$C_2 = |\mathrm{Cot}\,2\,(\Delta T + \alpha)|$$

$$\alpha = \Delta X_C\,|\,S1C2$$

where C_2 is represented by a line with a slope of cot 2 and $C_2 = 0$ for $\Delta T = -\alpha$. The minimum value of equation (6-49) must lie between $\Delta T = 0$ and $\Delta T = -\alpha$ but it may be:

(a) for $\Delta T = 0$, if (curve 1):
$\tan(\cot 2) > 1$ and $-\Delta X_C | S1C2 > 0$
or $0 > \tan(\cot 2) > -1$ and $-\Delta X_C | S1C2 < 0$

(b) for $\Delta T = -\alpha$, if (curve 2):
$0 < \tan(\cot 2) < 1$ and $-\alpha > 0$
or $\tan(\cot 2) < -1$ and $-\alpha < 0$

(c) for every value of ΔT between 0 and $-\alpha$, if (curve 3):
$\tan(\cot 2) = 1$ and $-\alpha > 0$
or $\tan(\cot 2) = -1$ and $-\alpha < 0$

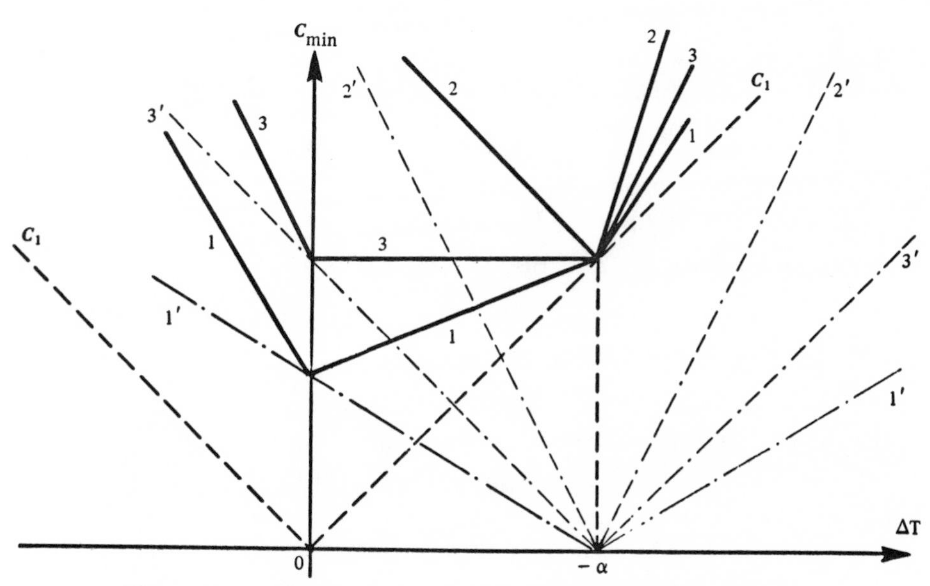

Figure 43. *Graphical representation of C_{min} as a function of ΔT*

According to the values of θ_2 and θ_1 and the sign of ΔX_C one of two solutions can be obtained:

$$\Delta\theta_1 = 0 \quad \Delta T = 0 \quad \Delta\theta_2 = \Delta X_C / S1S2(L_2 + T) \qquad \textbf{(6-50)}$$

$$\Delta\theta_1 = 0 \quad \Delta T = -\Delta X_C / S1C2 \quad \Delta\theta_2 = 0 \qquad \textbf{(6-51)}$$

In the case of curve 3, the two solutions lead to a minimization of the criterion. Apart from its practical significance, this criterion is easy to program and is of more practical interest than the previous quadratic criteria.

6.3.3 THE METHOD OF GENERALIZED INVERSES[20]

6.3.3.1 The principles involved

Consider, once again, equation (6-5):

$$\underset{n \times 1}{\underline{\Delta X}} = \underset{n \times m}{J} \; \underset{m \times 1}{\underline{\Delta \Theta}}$$

The generalized inverse of J is every matrix [G] which satisfies the equation:

$$JGJ = J \tag{6-52}$$

A solution of equation (6-5) can then be written as:

$$\underset{m \times 1}{\underline{\Delta \Theta}} = \underset{m \times n}{G} \; \underset{n \times 1}{\underline{\Delta X}} \tag{6-53}$$

There are m − n + 1 linearly independent solutions for $\underline{\Delta\Theta}$. If $\underline{\Delta\Theta_k}$ is one of these solutions, then every vector with the form:

$$\underline{\Delta\Theta_q} = \sum_{k=1}^{m-n+1} \lambda_k \underline{\Delta\Theta_k} \text{ with } \sum_{k=1}^{m-n+1} \lambda_k = 1 \tag{6-54}$$

is also a solution of equation (6-5).

Example: Apply this to equation (6-17):

$$\begin{pmatrix} \Delta X_C \\ \Delta Z_C \end{pmatrix} = \begin{pmatrix} -C1C2(L_2+T) & S1S2(L_2+T) & -S1C2 \\ 0 & C2\,(L_2+T) & S2 \end{pmatrix} \begin{pmatrix} \Delta\theta_1 \\ \Delta\theta_2 \\ \Delta T \end{pmatrix}$$

This matrix can be partitioned in three different ways, thus generating the square matrices $J\theta_2, J\theta_1, J_T$ [see equations (6-23) to (6-25)]:

$$J\theta_2 = \begin{pmatrix} -C1C2(L_2+T) & -S1C2 \\ 0 & S2 \end{pmatrix}$$

$$[J\theta_2]^{-1} = \frac{1}{C1C2S2(L_2+T)} \begin{pmatrix} -S2 & -S1C2 \\ 0 & C1C2(L_2+T) \end{pmatrix}$$

$$J\theta_1 = \begin{pmatrix} S1S2(L_2+T) & -S1C2 \\ C2(L_2+T) & S2 \end{pmatrix}$$

$$[J\theta_1]^{-1} = \frac{1}{S1(L_2+T)} \begin{pmatrix} S2 & S1C2 \\ -C2(L_2+T) & S1S2(L_2+T) \end{pmatrix}$$

$$J_T = \begin{pmatrix} -C1C2(L_2+T) & S1S2(L_2+T) \\ 0 & C2(L_2+T) \end{pmatrix}$$

$$[J_T]^{-1} = \frac{1}{C1C2^2(L_2+T)} \begin{pmatrix} -C2 & S1S2 \\ 0 & C1C2 \end{pmatrix}$$

The three generalized inverses of J are then:

$$[G_{\theta_2}] = \frac{-1}{C1C2S2(L_2+T)} \begin{pmatrix} S2 & S1S2 \\ 0 & 0 \\ 0 & -C1C2(L_2+T) \end{pmatrix}$$

$$[G_{\theta_1}] = \frac{1}{S1\,(L_2+T)} \begin{pmatrix} 0 & 0 \\ S2 & S1S2 \\ -C2(L_2+T) & S1S2(L_2+T) \end{pmatrix}$$

$$[G_T] = \frac{-1}{C1C2^2\,(L_2+T)} \begin{pmatrix} C2 & -S1S2 \\ 0 & -C1C2 \\ 0 & 0 \end{pmatrix}$$

It can be verified that for each of these matrices equation (6-52) is satisfied. They correspond to three solutions:

$$\underline{\Delta\Theta}_{\theta_2} \begin{cases} \Delta\theta_1 = -\Delta X_C/[C1C2(L_2+T)] - S1\Delta Z_C/[C1S2(L_2+T)] \\ \Delta\theta_2 = 0 \\ \Delta T = \Delta Z_C/S2 \end{cases}$$

$$\underline{\Delta\Theta}_{\theta_1} \begin{cases} \Delta\theta_1 = 0 \\ \Delta\theta_2 = [S2\Delta X_C/S1(L_2+T)]+[C2\,\Delta Z_C/(L_2+T)] \\ \Delta T = [-C2\,\Delta X_C/S1] + S2\,\Delta Z_C \end{cases}$$

$$\underline{\Delta\Theta}_T \begin{cases} \Delta\theta_1 = [-\Delta X_C/C1C2(L_2+T)] + S1S2\,\Delta Z_C/C1C2^2\,(L_2+T) \\ \Delta\theta_2 = \Delta Z_C/C2(L_2+T) \\ \Delta T = 0 \end{cases}$$

Any two of the three solutions $\underline{\Delta\Theta}_{\theta_1}$, $\underline{\Delta\Theta}_{\theta_2}$ and $\underline{\Delta\Theta}_T$ are linearly independent and may form a basis for the solution of equation (6-17).

Each linear combination satisfying equation (6-54) might be a possible solution. For arm control the choice is important since it dictates diverse computations and movements.

6.3.3.2 Another method of computation of the solutions

Rather than computing m − n + 1 independent solutions all possible solutions can more simply be determined by one computation of generalized inverse. If a solution of equation (6-5) is written as:

$$\underline{\Delta\Theta} = G\,\underline{\Delta X} \qquad \textbf{(6-53)}$$

then every value of $\underline{\Delta\Theta}^*$, such as:

$$\underline{\Delta\Theta}^* = \underline{\Delta\Theta} + (GJ - \Uparrow)\,\underline{z} \qquad \textbf{(6-55)}$$

is a solution of the equation. The vector $\underline{z}$ has m − n arbitrary coefficients which are linearly independent.

6.3.3.3 *Computation of generalized inverses*

The matrix J (n × m), being of rank n, can be partitioned if in the form:

$$J = [J_1, J_2]$$

where J is a square matrix of order n. The matrix $G = [J_1^{-1}\ 0]^T$ is a generalized inverse of J. This is the method used in the previous example and is identical to the method in which principal variables are used.

6.3.3.4 *An important generalized inverse: the pseudo-inverse J^+*

Since an infinite number of generalized inverses exist, it becomes possible to choose a solution which minimizes a criterion. The pseudo-inverse provides a solution which minimizes the Euclidean form of movements, $\sum_i (\Delta\theta_i)^2$ – the quadratic criterion mentioned previously.

An algorithm described by Greville[20, 21] allows J^+ to be determined quickly and without the need for inversion of the matrix. This algorithm is very useful as the inversion of a matrix using classical methods is problematic since the values used are zero or approaching zero. Division by zero or by ϵ does not allow further computation.

The Greville algorithm is derived as follows: let the Kth column of J and J_{K-1} of the submatrix constituted by the first K – 1 columns of J,

$$J_K = (J_{K-1}, \underline{j}_K)$$

The pseudo-inverse of J_K can be written as:

$$J_K^+ = \begin{pmatrix} J_{K-1}^+ - \underline{d}_K \underline{b}_K^T \\ \underline{b}_K^T \end{pmatrix} \quad (6\text{-}56)$$

with:

$$\underline{d}_K = J_{K-1}^+ \underline{j}_K \quad (6\text{-}57)$$

$$\underline{c}_K = \underline{j}_K - J_{K-1} \underline{d}_K \quad (6\text{-}58)$$

$$\text{if } \underline{c}_K \neq 0 \quad \underline{b}_K^T = \underline{c}_K^+ = \underline{c}_K^T / \underline{c}_K^T \underline{c}_K \quad (6\text{-}59)$$

$$\text{if } \underline{c}_K = 0 \quad \underline{b}_K^T = \underline{d}_K^T \cdot J_{K-1}^+ / (1 + \underline{d}_K^T \underline{d}_K) \quad (6\text{-}60)$$

The algorithm is therefore iterative. The initialization of the iterations depends on the first column of J:

$$\text{if } \underline{j}_1 = 0 \text{ then } J_1^+ = \underline{0}^T \quad (6\text{-}61)$$

$$\text{if } \underline{j}_1 \neq 0 \quad J_1^+ = \underline{j}_1^T / (\underline{j}_1^T \underline{j}_1) \quad (6\text{-}62)$$

The algorithm is independent of the rank and dimensions of J. It allows on-line detection of the singularities of J at each iteration, by computation of the J + J trace which is equal to the rank of matrix J and of matrix J^+.

Example: Consider equations (6-9) and (6-10):

$$\Delta X_C = -C1C2L\Delta\theta_1 + S1S2L\Delta\theta_2 - S1C2\Delta T \quad \textbf{(6-9)}$$

$$\Delta Y_C = -S1C2L\Delta\theta_1 - C1S2L\Delta\theta_2 + C1C2\Delta T \quad \textbf{(6-10)}$$

Matrix J can be written as:

$$J = \begin{pmatrix} -C1C2L & S1S2L & -S1C2 \\ -S1C2L & -C1S2L & C1C2 \end{pmatrix} \quad \textbf{(6-63)}$$

Applying Greville's algorithm:

$$\begin{cases} J_3 = (J_2, \underline{j}_3) \\ J_3^+ = \begin{pmatrix} J_2^+ - \underline{d}_3\, \underline{b}_3^T \\ \underline{b}_3^T \end{pmatrix} \\ \underline{d}_3 = J_2^+ \underline{j}_3 \\ \underline{c}_3 = \underline{j}_3 - J_2\, \underline{d}_3 \end{cases} \qquad \begin{cases} J_2 = (J_1, \underline{j}_2) \\ J_2^+ = \begin{pmatrix} J_1^+ - \underline{d}_2\, \underline{b}_2^T \\ \underline{b}_2^T \end{pmatrix} \\ \underline{d}_2 = J_1^+ \underline{j}_2 \\ \underline{c}_2 = \underline{j}_2 - J_1\, \underline{d}_2 \end{cases}$$

$$\begin{cases} J_1 = \underline{j}_1 = \begin{pmatrix} -C1C2L \\ -S1C2L \end{pmatrix} \\ \text{suppose that } J_1 \neq 0 \\ J_1^+ = \underline{j}_1^T / \underline{j}_1^T \underline{j}_1 = (-C1/C2L, -S1/C2L) \end{cases}$$

$$\begin{cases} \underline{d}_2 = (-C1/C2L\ -S1/C2L) \begin{pmatrix} S1S2L \\ -C1S2L \end{pmatrix} = 0 \\ \underline{c}_2 = \underline{j}_2 = \begin{pmatrix} S1S2L \\ -C1S2L \end{pmatrix} \\ \underline{b}_2^T = \underline{c}_2^T / \underline{c}_2^T \underline{c}_2 = \underline{j}_2^T / \underline{j}_2^T \underline{j}_2 = (S1/S2L, -C1/S2L) \\ J_2^+ = \begin{pmatrix} -C1/C2L & -S1/C2L \\ S1/S2L & -C1/S2L \end{pmatrix} \end{cases}$$

$$\begin{cases} \underline{d}_3 = J_2^+ \begin{pmatrix} -S1C2 \\ C1C2 \end{pmatrix} = \begin{pmatrix} 0 \\ -C2/S2L \end{pmatrix} \\ \underline{c}_3 = \begin{pmatrix} -S1C2 \\ C1C2 \end{pmatrix} - \begin{pmatrix} -C1C2L & S1S2L \\ -S1C2L & -C1S2L \end{pmatrix} \begin{pmatrix} 0 \\ -C2/S2L \end{pmatrix} = \begin{pmatrix} 0 \\ 0 \end{pmatrix} \\ \underline{b}_3^T = \underline{d}_3^T J_2^+ / (1 + \underline{d}_3^T \underline{d}_3) = [-S1C2/(C2^2 + S2^2 L^2)\quad C1C2/(C2^2 + S2^2 L^2)] \end{cases}$$

which gives:

$$J_3^+ = \begin{pmatrix} -C1/C2L & -S1/C2L \\ S1S2L/(C2^2+S2^2L^2) & -C1S2L/(C2^2+S2^2L^2) \\ -S1C2/(C2^2+S2^2L^2) & C1C2/(C2^2+S2^2L^2) \end{pmatrix}$$

The solution minimizing $(\Delta\theta_1)^2 + (\Delta\theta_2)^2 + (\Delta T)^2$ is then:

$$\Delta\theta_1 = (-C1/C2L)\Delta X_C - (S1/C2L)\Delta Y_C$$

$$\Delta\theta = [S1S2L/(C2^2+S2^2L^2)]\Delta X_C - [C1S2L/(C2^2+S2^2L^2)]\Delta Y_C$$

$$\Delta T = [-S1C2/(C2^2+S2^2L^2)]\Delta X_C + [C1C2/(C2^2+S2^2L^2)]\Delta Y_C$$

The only restriction, detectable on-line, and which itself modifies the development of the algorithm, is the singularity of $\theta_2 = \pi/2$. J is then equal to:

$$\begin{pmatrix} 0 & S1L & 0 \\ 0 & -C1L & 0 \end{pmatrix}$$

and a relation between ΔX_C and ΔY_C is obtained (see section 6.2).

6.3.3.5 Optimization of a second criterion using the pseudo-inverse

Choosing as the generalized inverse the solution $G = J^+$, equation (6-55) becomes:

$$\underline{\Delta\Theta}^* + J^+\underline{\Delta X} + (J^+J - 1\!\!1)\,\underline{z} \qquad \textbf{(6-64)}$$

The first part of the right-hand side of the equation minimizes the Euclidian norm of arm displacements. In the second part, $\underline{z}$ can be specified as a vector, having arbitrary components. It can be considered as the gradient of a positive, definite scalar function, depending on the state $\underline{\Theta}$ of the system. It represents a cost function.

Therefore, in addition to minimizing the Euclidian norm of displacements, the use of equation (6-64) allows costs to be minimized.

Fournier,[22] for example, used equation (6-64) for the automatic control of a manipulator with six DOF by taking the distance of the articulations from their mean positions as a cost function to avoid blocking the arms during manipulation.

$$\phi(\underline{\Theta}) = \frac{1}{6}\sum_{i=1}^{6}\left(\frac{\theta_i - \theta_{i\,average}}{\theta_{i\,max} - \theta_{i\,average}}\right)^2$$

$\underline{z}$ is then the gradient of ϕ.

Conclusions

The use of equation (6-64) provides a solution for the problem of speed control in articulated robots. This allows:

(a) a 'double' optimization: (i) a 'forced' optimization which corresponds to a useful criterion, ie minimizing the displacement of a set of articulations; and (ii) an 'optional' optimization which allows the criterion most suitable for a particular application to be determined.
(b) a valid solution, whatever the dimensions of $\underline{\Delta X}$ and $\underline{\Delta \Theta}$.
(c) the use of a computer – using algorithms such as the one described by Greville, there is no need for matrix inversion and the singularities can be detected on-line and, therefore, the computer will not 'block' the control.
(d) the time taken for computation to be reduced (usually to a few milliseconds, depending on the number of DOF).

Chapter 7

Articulated mechanical systems: the dynamic model

A characteristic inherent in all models used in automatic control systems is that they are valid only for a limited range of applications. For positional and speed control of AMS, geometrical and kinematic models are used which assume, for any configuration taken up by the robot, that a state of equilibrium is achieved. It is clear, therefore, that these models become less representative as the speed increases and inertial, centrifugal and coupling forces become significant (and also friction and elasticity come into effect). Because of the high speeds required on robot-operated production lines, certain dynamic phenomena should be considered before deciding on a model.

7.1 A dynamic model for an open articulated chain of rigid segments, without backlash or friction

7.1.1 NOTATION

There are several possible sources for such a model: Newton-Euler's equation,[23, 24] Gibbs' functions,[25] d'Alembert's formalisms,[26] bond graphs,[27, 28] and Lagrange equations.[29, 30] The equations described by Lagrange are used most frequently because they are easy to compute.

Consider a chain composed of N articulated segments, 1 to N, with segment 0 being fixed. According to Lagrange, a representation of the movment of the chain can be written (assuming the points discussed in section 3.1 hold true) as:

$$\begin{cases} \dfrac{d}{dt}\left(\dfrac{\partial L}{\partial \dot{\theta}_i}\right) - \dfrac{\partial L}{\partial \theta_i} = Q_i + \Gamma_i \\ i = 1, N \end{cases} \tag{7-1}$$

The Lagrange factor (L) expresses the difference between the kinetic energy (T) and the potential energy (U) of the system. Since, in the present case, the segments of the chain are rigid, L can be reduced to the kinetic energy, which is the sum of the individual energies (T_λ) of N segments (C_λ) arranged as a chain. Equation (7-1) can then be written as:

$$\left\{\begin{array}{l}\sum\limits_{\lambda=1}^{N} \dfrac{d}{dt}\left(\dfrac{\partial T_\lambda}{\partial \dot{\theta}_i}\right) - \dfrac{\partial T_\lambda}{\partial \theta_i} = Q_i + \Gamma_i \\ i = 1, N\end{array}\right. \tag{7-2}$$

where θ_i is the first independent variable of the system (often the angle between two consecutive segments or the translation between these two segments), and:

$$\dot{\theta}_i = d\theta_i/dt$$

Q_i is the torque exerted by gravity on the ith articulated variable, and Γ_i is the torque exerted by the outside forces (often due to the actuators) on the ith articulation.

7.1.2 MATRIX REPRESENTATION

The kinetic energy (T_λ) of a segment (C_λ) is the sum of its kinetic energy of translation [the mass (M_λ) of the body is assumed to be concentrated at its mass centre (G_λ)] and its kinetic energy of rotation about an axis (z) which passes through its mass centre (see Figure 44):

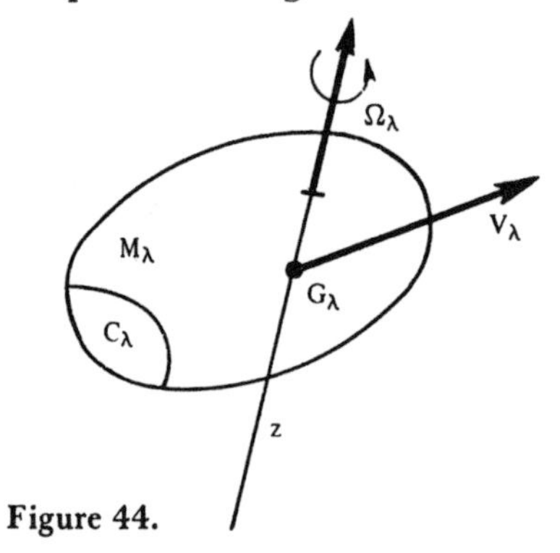

Figure 44.

Using the notations described in Chapter 3, the kinetic energy of the segment illustrated in Figure 44 can be written as:

$$2T_\lambda = M_\lambda\,[\underline{V}^{(0)}_{G_\lambda}(R_\lambda)]^2 + [\underline{\Omega}^{(0)}_\lambda(R_\lambda)]^T\,\underline{\underline{I}}^\lambda\,[\underline{\Omega}^{(0)}_\lambda(R_\lambda)] \tag{7-3}$$

where $\underline{\underline{I}}_\lambda$ is the tensor of inertia of M_λ about z. Because the speed is determined with respect to a fixed set of coordinate axes (R_o) and expressed in the set of coordinate axes bound to C_λ, an abridged notation of equation (7-3) can be written:

$$2\,T_\lambda = M_\lambda\,\underline{V}^2_\lambda + \underline{\Omega}^T_\lambda\,\underline{\underline{I}}^\lambda\,\underline{\Omega}_\lambda \tag{7-4}$$

With reference to sections 3.6.1 and 3.5, it is clear that $\underline{V}^{(0)}_{G_\lambda}$ (R_λ) and $\underline{\Omega}^{(0)}_\lambda$ (R_λ) have the same characteristics. These speeds are expressed by a vector having three components, each with the form:

$$\sum_{i=1}^{N} F_i(\theta_1 \ldots \theta_N) . \dot{\theta}_i$$

Suppose that:

$$\underline{V}_{G\lambda}^{(0)}(R_\lambda) \text{ where } \underline{\Omega}_\lambda^{(0)}(R_\lambda) = \begin{bmatrix} \sum_{i=1}^{N} F_i^x(\theta_1 \ldots \theta_N) . \dot{\theta}_i \\ \sum_{i=1}^{N} F_i^y(\theta_1 \ldots \theta_N) . \dot{\theta}_i \\ \sum_{i=1}^{N} F_i^z(\theta_1 \ldots \theta_N) . \dot{\theta}_i \end{bmatrix} \tag{7-5}$$

the kinetic energy of C_λ will, in the light of equation (7-3), take the form:

$$T_\lambda = \sum_{i=1}^{N} \sum_{j=1}^{N} g_{ij}(\theta_1 \ldots \theta_N) . \dot{\theta}_i \dot{\theta}_j \tag{7-6}$$

The constituent parts of equation (7-2) can then be written:

$$\frac{\partial T_\lambda}{\partial \theta_k} = \sum_{i=1}^{N} \sum_{j=1}^{N} \frac{\partial g_{ij}(\theta_1 \ldots \theta_N)}{\partial \theta_k} . \dot{\theta}_i \dot{\theta}_j \tag{7-7}$$

$$\frac{\partial T_\lambda}{\partial \dot{\theta}_k} = \sum_{j=1}^{N} g_{kj}(\theta_1 \ldots \theta_N) . \dot{\theta}_j \tag{7-8}$$

$$\frac{d}{dt}\left(\frac{\partial T_\lambda}{\partial \dot{\theta}_k}\right) = \sum_{j=1}^{N} g_{kj}(\theta_1 \ldots \theta_N) . \ddot{\theta}_j + \sum_{j=1}^{N} \frac{\partial g_{kj}(\theta_1 \ldots \theta_N)}{\partial \theta_k} . \dot{\theta}_k \dot{\theta}_j \tag{7-9}$$

C_λ in the left-hand side of equation (7-2) (relative to the independent variable, θ_k) provides terms in θ_j (j from 1 to N) and terms in $\theta_i\ \theta_j$ (i and j from 1 to N). Considering the contribution made by all the segments, the following equation (corresponding to the variable θ_k) can be written:

$$\underline{a}_k^T(\theta_1 \ldots \theta_N) . \underline{\ddot{\Theta}} + \underline{b}_k^T(\theta_1 \ldots \theta_N) . \underline{\dot{\Theta}}^2 + \underline{c}_k^T(\theta_1 \ldots \theta_N) . \underline{\dot{\Theta}\dot{\Theta}} = Q_k + \Gamma_k \tag{7-10}$$

where $\underline{a_k}$ is a vector containing elements g_{kj} (dimension N), $\underline{b_k}$ is a vector containing elements $\partial g_{ii}/\partial \theta_k$ (dimension N), $\underline{c_k}$ is a vector with C_N^2 components $\partial g_{ii}/\partial \theta_k$ ($i \neq j$), $\underline{\ddot{\Theta}}$ represents the vector $(\ddot{\theta}_1 \ddot{\theta}_2 \ldots \ddot{\theta}_N)^T$, $\underline{\dot{\Theta}}^2$ represents the vector $(\dot{\theta}_1^2 \dot{\theta}_2^2 \ldots \dot{\theta}_N^2)^T$, and $\underline{\dot{\Theta}\dot{\Theta}}$ represents the vector $(\dot{\theta}_1\dot{\theta}_2, \dot{\theta}_1\dot{\theta}_3, \ldots \dot{\theta}_1\dot{\theta}_N, \dot{\theta}_2\dot{\theta}_3, \ldots \dot{\theta}_2\dot{\theta}_N, \ldots \dot{\theta}_{N-1}\dot{\theta}_N)^T$.

Equation (7-2) can then be described by N equations [of a type similar to equation (7-10)] which can easily be expressed in matrix form:

$$\underset{N \times N}{[A]}\ \underset{N \times 1}{\underline{\ddot{\Theta}}} + \underset{N \times N}{[B]}\ \underset{N \times 1}{\underline{\dot{\Theta}}^2} + \underset{N \times C_N^2}{[C]}\ \underset{C_N^2 \times 1}{\underline{\dot{\Theta}\dot{\Theta}}} = \underset{N \times 1}{\underline{Q}(\underline{\Theta})} + \underset{N \times 1}{\underline{\Gamma_\Theta}} \tag{7-11}$$

The matrices [A], [B] and [C] and vector $\underline{Q}$ have elements which are functions of variables θ_1 to θ_N. The terms in matrices A, B and C can be referred to as the *dynamic coefficients* of the system. Equation (7-11) can be considered as N, non-linear, coupled second-order differential equations. Alternatively, equation (7-11) can be presented as:

$$\left\{ \begin{array}{l} \sum_{j=1}^{N} \left\{ A(i,j)\ddot{\theta}_j + B(i,j)\dot{\theta}_j^2 + \sum_{k=j+1}^{N} C(i,j,k)\dot{\theta}_j\dot{\theta}_k \right\} = Q_i + \Gamma_{\theta_i} \\ i = 1, N \end{array} \right. \tag{7-12}$$

Equations (7-11) and (7-12) indicate clearly the different forces acting on the system. Thus, $[A]\ddot{\underline{\Theta}}$ corresponds to the inertial forces, [A] being the coefficient matrix of the forces due to acceleration; [B] the coefficient matrix of the centrifugal forces; [C] the coefficient matrix of the coupling Coriolis forces; $\underline{Q}(\underline{\Theta})$ the vector of the gravitational forces; and $\underline{\Gamma}_\theta$ the vector of the external forces applied to the system.

7.1.3 COMPUTATION OF DYNAMIC COEFFICIENTS

To determine vectors $\underline{a}_k$, $\underline{b}_k$ and $\underline{c}_k$ (relative to a variable, θ_k) for each of the segments in a chain, the equations described in Chapter 3, which were used to compute the speed equations (7-6) to (7-9), can be used. By repeating this procedure for other variables (θ_i), A(i,j), B(i,j), and C(i,j,k) can be determined. Programs have been written which, starting from an adequate description of the system (see Chapter 2), will automatically compute the dynamic coefficients. Three such programs are TOAD,[31] OSSAM[32] and EDYLIMA.[33]

7.1.4 DETERMINATION OF $\underline{Q}$, THE VECTOR OF THE GRAVITATIONAL FORCES

Suppose that the variable θ_i is increased by an amount $\Delta\theta_i$, then the centres of gravity of the various segments below θ_i will be at a height $Y^{(0)}_{G_\lambda}$, incremented by $\Delta Y^{(0)}_{G_\lambda}$. By applying d'Alembert's principle (the principle of virtual effect) it can be postulated that the torque (Q_i), acting on θ_i, during a movement of $\Delta\theta_i$ does work equal to, and corresponding to, a vertical movement of the centre of gravity (G_λ) of the segments (C_λ) of mass, M_λ. Then:

$$Q_i\Delta\theta_i = \sum_{\lambda=1}^{N} gM_\lambda \, \Delta Y^{(0)}_{G_\lambda} \tag{7-13}$$

which leads to:

$$Q_i = \sum_{\lambda=1}^{N} gM_\lambda \frac{\partial Y^{(0)}_{G_\lambda}}{\partial \Theta_i} \tag{7-14}$$

where g is the acceleration due to gravity.

7.2 Development of a dynamic equation for a system having three degrees of freedom

How dynamic equations are developed can be illustrated by considering the arm shown in Figure 28 (which involves two rotations followed by a translation) which is reproduced in Figure 45 using notations better suited to the writing of dynamic equations.

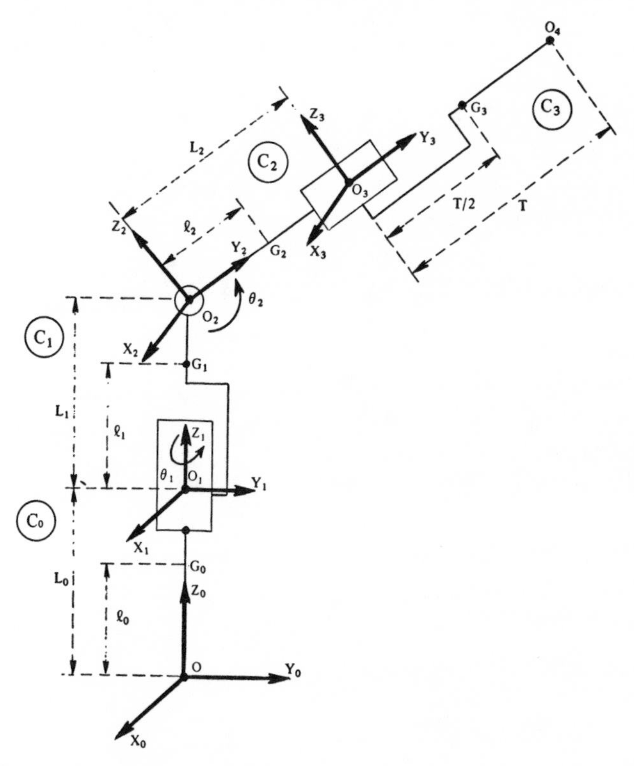

Figure 45a. *Articulation of a limb involving two rotations and one translation, and its associated set of coordinate axes*

First the kinetic energy of each segment (C_0, C_1, C_2 and C_3 with masses M_0, M_1, M_2 and M_3, and tensors of inertia* $\underline{\underline{I}}^{(0)}$, $\underline{\underline{I}}^{(1)}$, $\underline{\underline{I}}^{(2)}$ and $\underline{\underline{I}}^{(3)}$ about their axes of rotation) must be determined:

$$2T_\lambda = M_\lambda \, [\underline{V}^{(0)}_{G_\lambda}(R_\lambda)]^2 + \underline{\Omega}^{(0)T}_\lambda (R_\lambda) \, \underline{\underline{I}}^{(\lambda)} \, \underline{\Omega}^{(0)}_\lambda (R_\lambda) \tag{7-3}$$

* The tensors of inertia are expressed thus:

$$\underline{\underline{I}}^{(k)} = \begin{bmatrix} I^{(k)}_{11} & I^{(k)}_{12} & I^{(k)}_{13} \\ I^{(k)}_{21} & I^{(k)}_{22} & I^{(k)}_{23} \\ I^{(k)}_{31} & I^{(k)}_{32} & I^{(k)}_{33} \end{bmatrix}$$

This computation requires the evaluation of speeds and coordinate transformations by the methods described in Chapter 3.

The following formulae can be used:

(a) coordinate transformation matrices:

$$M_0^1 = \begin{pmatrix} C1 & -S1 & 0 \\ S1 & C1 & 0 \\ 0 & 0 & 1 \end{pmatrix} \quad M_1^2 = \begin{pmatrix} 1 & 0 & 0 \\ 0 & C2 & -S2 \\ 0 & S2 & C2 \end{pmatrix} \quad M_2^3 = (\Uparrow)$$

(b) transformation of the rotational velocity of a set of coordinate axes:

$$\underline{\Omega}_\lambda^{(0)}(R_\lambda) = \overset{+}{\underline{\Omega}}_\lambda^{\lambda-1}(R_{\lambda-1}) + [M_{\lambda-1}^{\lambda}]^T \, \underline{\Omega}_{\lambda-1}^{(0)}(R_{\lambda-1}) \quad \text{(3-22)}$$

(c) transformation of the translational velocity of a set of coordinate axes when the DOF are rotational:

$$\underline{V}_{O_{\lambda+1}}^{(0)}(R_\lambda) = \underline{V}_{O_\lambda}^{(0)}(R_\lambda) + \underline{\Omega}_\lambda^{(0)}(R_\lambda) \wedge \underline{O_\lambda O_{\lambda+1}}(R_\lambda) \quad \text{(3-26)}$$

$$\underline{V}_{O_{\lambda+1}}^{(0)}(R_\lambda) = [M_\lambda^{\lambda+1}]\, \underline{V}_{O_{\lambda+1}}^{(0)}(R_{\lambda+1}) \quad \text{(3-27)}$$

(d) transformation of the translational velocity of a set of coordinate axes when the DOF is translational:

$$\underline{V}_{O_{\lambda+1}}^{(p)}(R_\lambda) = \underline{V}_{O_\lambda}^{(p)}(R_\lambda) + \underline{\Omega}_\lambda^{(p)}(R_\lambda) \wedge \underline{O_\lambda O_{\lambda+1}}(R_\lambda) + \underline{V}_{O_{\lambda+1}}^{O_\lambda}(R_\lambda) \quad \text{(3-28)}$$

(e) rotational velocity $\overset{+}{\underline{\Omega}}_\lambda^{\lambda-1}(R_{\lambda-1})$:

$$\overset{+}{\underline{\Omega}}_1^0 (R_0) = [0, 0, \dot{\theta}_1]^T$$
$$\overset{+}{\underline{\Omega}}_2^1 (R_1) = [\dot{\theta}_2, 0, 0]^T$$
$$\overset{+}{\underline{\Omega}}_3^2 (R_2) = [0, 0, 0]^T$$

For each segment, equation (7-3) will have the successive values of:

$$2\,T_0 = 0$$

$$2\,T_1 = M_1 \, [\underline{V}_{G_1}^{(0)}(R_1)]^2 + \underline{\Omega}_1^{(0)T}(R_1)\, \underline{\underline{I}}^{(1)} \underline{\Omega}_1^{(0)}(R_1)$$

$$\underline{V}_{G_1}^{(0)}(R_1) = 0$$

$$\underline{\Omega}_1^{(0)}(R_1) = \overset{+}{\underline{\Omega}}_1(R_0) + [M_0^1]^T \, \underline{\Omega}_0^{(0)}(R_0) = [0, 0, \dot{\theta}_1]^T$$

$$2\,T_1 = \overset{(1)}{I_{33}} \,.\, \dot{\theta}_1^2$$

$$2\,T_2 = M_2\,[\underline{V}^{(0)}_{G_2}(R_2)]^2 + \underline{\Omega}_2^{(0)T}(R_2)\,\underline{\underline{I}}^{(2)}\,\underline{\Omega}_2^{(0)}(R_2)$$

$$\underline{\Omega}_2^{(0)}(R_2) = \overset{+}{\underline{\Omega}}{}_2^1(R_1) + [M_1^2]^T\,\underline{\Omega}_1^{(0)}(R_1)$$

$$= [\dot{\theta}_2, S2\,\dot{\theta}_1, C2\,\dot{\theta}_1]^T$$

$$\underline{V}^{(0)}_{G_2}(R_2) = \underline{V}^{(0)}_{O_2}(R_2) + \underline{\Omega}_2^{(0)}(R_2) \wedge \underline{O_2\,G_2}\,(R_2)$$

$$= [S2\,\ell_2\,\dot{\theta}_1, \ell_2\,\dot{\theta}_2, 0]^T$$

$$2\,T_2 = \dot{\theta}_2^2\,(M_2\,\ell_2^2 + \overset{(2)}{I_{11}}) + \dot{\theta}_1^2\,[M_2\,\ell_2^2\,S2^2 + \overset{(2)}{I_{22}}\,S2^2 + \overset{(2)}{I_{33}}\,C2^2 + S2C2(\overset{(2)}{I_{32}} + \overset{(2)}{I_{23}})] + \dot{\theta}_1\,\dot{\theta}_2\,[C2(\overset{(2)}{I_{13}} + \overset{(2)}{I_{31}}) + S2(\overset{(2)}{I_{12}} + \overset{(2)}{I_{21}})]$$

$$2\,T_3 = M_3\,[\underline{V}^{(0)}_{G_3}(R_3)]^2 + \underline{\Omega}_3^{(0)}(R_3)^T\,\underline{\underline{I}}^{(3)}\,\underline{\Omega}_3^{(0)}(R_3)$$

$$\underline{\Omega}_3^{(0)}(R_3) = \overset{+}{\underline{\Omega}}{}_3^2(R_2) + [M_2^3]^T\,\underline{\Omega}_2^{(0)}(R_2)$$

$$= [\dot{\theta}_2, S2\dot{\theta}_1, C2\dot{\theta}_1]^T$$

$$\underline{V}^{(0)}_{G_3}(R_3) = \underline{V}^{(0)}_{O_3}(R_3) + \underline{\Omega}_3^{(0)}(R_3) \wedge \underline{O_3\,G_3}\,(R_3) + \underline{V}^{O_3}_{G_3}(R_3)$$

$$= [-\,S2\,\dot{\theta}_1\,T/2, \dot{\theta}_2\,T/2, \dot{T}]$$

$$2\,T_3 = M_3\,\dot{T}^2 + \dot{\theta}_1^2\,[(M_3\,T^2\,S2^2/4) + \overset{(3)}{I_{22}}\,S2^2 + \overset{(3)}{I_{33}}\,C2^2 + S2C2\,(\overset{(3)}{I_{23}} + \overset{(3)}{I_{32}})] + \dot{\theta}_2^2\,[(M_3\,T^2/4) + \overset{(3)}{I_{11}}] + \dot{\theta}_1\,\dot{\theta}_2\,[S2\,(\overset{(3)}{I_{12}} + \overset{(3)}{I_{21}}) + C2\,(\overset{(3)}{I_{13}} + \overset{(3)}{I_{31}})]$$

The Lagrangian is then expressed by the equation:

$$L = \sum_{\lambda=0}^{3} 2\,T_\lambda = \dot{\theta}_1^2\,[\overset{(1)}{I_{33}} + \overset{(2)}{I_{22}}\,S2^2 + \overset{(2)}{I_{33}}\,C2^2 + M_2\,\ell_2^2\,S2^2 + \overset{(2)}{I_{22}}\,S2^2 + \overset{(3)}{I_{33}}\,C2^2 + M_3 T^2\,S2^2/4] + \dot{\theta}_2^2\,[M_2\,\ell_2^2 + \overset{(2)}{I_{11}} + \overset{(3)}{I_{11}} + M_3 T^2/4] + M_3\,\dot{T}^2 + \dot{\theta}_1\,\dot{\theta}_2\,[S2\,(\overset{(2)}{I_{12}} + \overset{(3)}{I_{12}} + \overset{(2)}{I_{21}} + \overset{(3)}{I_{21}}) + C2\,(\overset{(3)}{I_{13}} + \overset{(3)}{I_{13}} + \overset{(2)}{I_{31}} + \overset{(3)}{I_{31}})] \qquad \textbf{(7-15)}$$

Equation (7-15) can be simplified in the following way:

$$A = \overset{(1)}{I_{33}} \qquad D = M_2\,\ell_2^2 + \overset{(2)}{I_{11}} + \overset{(3)}{I_{11}} \qquad G = M_3/4$$

$$B = \overset{(2)}{I_{22}} + M_2\,\ell_2^2 + \overset{(3)}{I_{22}} \qquad E = \overset{(2)}{I_{12}} + \overset{(3)}{I_{12}} + \overset{(2)}{I_{21}} + \overset{(3)}{I_{21}}$$

$$C = \overset{(2)}{I_{33}} + \overset{(3)}{I_{33}} \qquad F = \overset{(2)}{I_{13}} + \overset{(3)}{I_{13}} + \overset{(2)}{I_{31}} + \overset{(3)}{I_{31}}$$

It is then possible to write equation (7-15) as:

$$L = \dot{\theta}_1^2\,[A + B\,S2^2 + CC2^2 + GT^2\,S2^2] + \dot{\theta}_2^2\,[D + GT^2] + \dot{\theta}_1\,\dot{\theta}_2\,[E\,S2 + FC2] + 4\,G\dot{T}^2$$

(f) calculation of the gravitational torque using equation (7-14):

$$Y_{G_1}^{(0)} = L_0 + \ell_1 \quad Y_{G_2}^{(0)} = L_0 + L_1 + \ell_2\, C2 \quad Y_{G_0}^{(0)} = \ell_0$$
$$Y_{G_3}^{(0)} = L_0 + L_1 + C2\,(L_2 + T/2) \qquad (7\text{-}16)$$

and

$$Q_0 = 0 \quad Q_1 = 0 \quad Q_2 = -gM_2\ell_2 S2 - gM_3 S2\,(L_2 + T/2)$$
$$Q_3 = gM_3 C2/2$$

The system of equation (7-2) can then be written as:

$$2\,\ddot{\theta}_1\,(A+BS2^2+CC2^2+GT^2S2^2) + \ddot{\theta}_2\,(ES2 + FC2) + \dot{\theta}_2^2\,(EC2 - FS2)$$
$$+ 2\,\dot{\theta}_1\,\dot{\theta}_2\,S2C2\,(2B - 2C + 2GT^2) + 4\,GS2^2T\dot{T}\dot{\theta}_1 = 2\,\Gamma_1 \qquad (7\text{-}17)$$
$$\ddot{\theta}_1\,(ES2 + FC2) + 2\,\ddot{\theta}_2\,(D + GT^2) - 2\,\dot{\theta}_1^2\,S2C2\,(B - C + GT^2) +$$
$$4\,\dot{\theta}_2\,\dot{T}T = -\,S2\,[2\,gM_2\,\ell_2 + gM_3\,(2L_2 + T)] + 2\,\Gamma_2 \qquad (7\text{-}18)$$
$$2\,M_3\,\ddot{T} - 2\,G\,S2^2\,\dot{\theta}_1^2 - 2\,T\,\dot{\theta}_2^2 = g\,M_3\,C2 + 2\,\Gamma_3 \qquad (7\text{-}19)$$

or, as a matrix:

$$[A]\begin{pmatrix}\ddot{\theta}_1\\ \ddot{\theta}_2\\ \ddot{T}\end{pmatrix} + [B]\begin{pmatrix}\dot{\theta}_1^2\\ \dot{\theta}_2^2\\ \dot{T}^2\end{pmatrix} + [C]\begin{pmatrix}\dot{\theta}_1\,\dot{\theta}_2\\ \dot{\theta}_1\,\dot{T}\\ \dot{\theta}_2\,\dot{T}\end{pmatrix} = \begin{pmatrix}2\Gamma_1\\ -S2[2gM_2\ell_2+gM_3(2L_2+T)]+2\Gamma_2\\ gM_3C2+2\Gamma_3\end{pmatrix} \qquad (7\text{-}20)$$

with:

$$[A] = \begin{pmatrix}2[A+BS2^2+CC2^2+GT^2S2^2] & ES2+FC2 & 0\\ ES2+FC2 & 2(D+GT^2) & 0\\ 0 & 0 & 2M_3\end{pmatrix}$$

$$[B] = \begin{pmatrix}0 & EC2-FS2 & 0\\ -2S2C2(B-C+GT^2) & 0 & 0\\ -2GS2^2 & -2T & 0\end{pmatrix}$$

$$[C] = \begin{pmatrix}4S2C2(B-C+GT^2) & 4GTS2^2 & 0\\ 0 & 0 & 4T\\ 0 & 0 & 0\end{pmatrix}$$

7.3 Another type of model: the bond graph

The bond graph, as proposed by Ezetial and Paynter,[27] is a graph which is constructed according to fundamental physical considerations of the storage, supply, dissipation and energy exchange of a system. The system to be studied is usually divided into two subsystems between which energy is exchanged via identifiable links or connections. Generation of a bond graph relies on the use of four generalized variables: effort (e), flux (f), displacement (q), and impusle (p) — all of which have an electrical significance, as well as being significant in hydraulics and mechanics.[34] In this last case effort is the force or torque, flux is the translational or rotational value, and impulse can be translational or rotational.

7.3.1 PARTS OF THE GRAPH

The symbols used in bond graphs are presented in Table A.

Symbol	Meaning
$\frac{e}{f}$—	single link
$\frac{e}{f}\rightharpoonup$ R	resistive element e = Rf or F(f)
$\frac{e}{f}\rightharpoonup$ I	inertial element $f = \frac{1}{I}\int edt$
$\frac{e}{f}\rightharpoonup$ C	capacitive element $e = \frac{1}{C}\int fdt$
Se—	source of effort
Sf—	source of flux
⟶	activation link

Symbol	Meaning
one-port elements	
– TF –	transformer $e_2 = n\,e_1$ $f_1 = n\,f_2$
– GY –	gyrator $e_2 = m\,f_1$ $e_1 = m\,f_2$
two-port elements	
$\frac{e_1}{f_1}$—0—$\frac{e_2}{f_2}$, with branch $e_3 \mid f_3$	parallel bond $e_1 = e_2 = e_3$ $f_1 + f_2 + f_3 = 0$
$\frac{e_1}{f_1}$—1—$\frac{e_2}{f_2}$, with branch $e_3 \mid f_3$	series bond $f_1 = f_2 = f_3$ $e_1 + e_2 + e_3 = 0$

Table A. *Symbols used in bond graphs*

Standard 'one-port' elements exchange energy with the system through a single route [restrictive elements (friction), inertial capacitive (springs)].

Standard 'two-port' elements exchange energy with the system through two linked elements [transformers (reducers), gyrators (gyroscopes)].

Standard 'bonding' elements (in parallel and in series); the energy flow is indicated by a half-arrow parallel with the bond, a short dash perpendicular to the edge of a bond indicating that the force is directed towards this end and the flow towards the other.

7.3.2 TREATMENT OF A SIMPLE EXAMPLE[34]

Consider the articulated system, possessing two DOF, which is illustrated in Figure 45b. The bond graph which is associated with this mechanism is shown in Figure 45c.

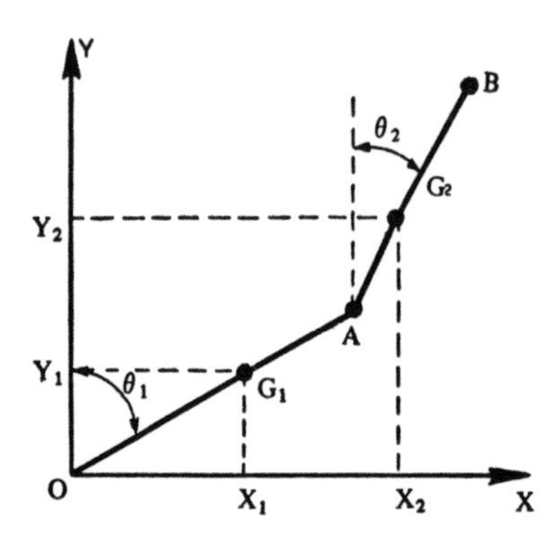

Figure 45b. *A mechanism possessing two DOF: $OG_1 = L_1$; $AG_2 = L_2$; $OA = \ell_1$ and $AB = \ell_2$ – its inertias and masses being j_1, j_2 and M_1, M_2, respectively*

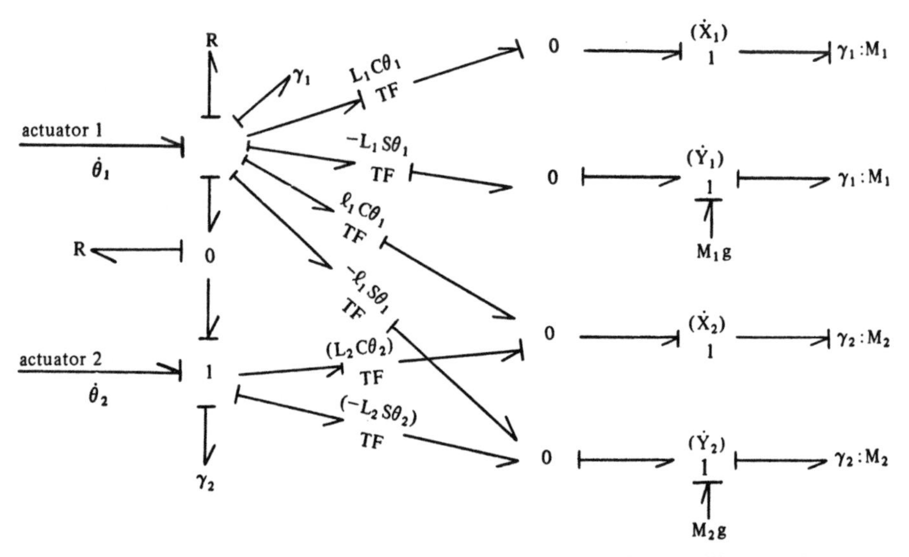

Figure 45c. *The bond graph associated with the mechanism illustrated in Figure 45b*

This bond graph was derived by consideration of the geometrical and speed transformations: the generalized variables being θ_1 and θ_2 and the geometrical variables X_1, Y_1, X_2 and Y_2. The following equations can be written:

$$X_1 = L_1 S1 \qquad Y_1 = L_1 C1$$
$$X_2 = \ell_1 S1 + L_2 S2 \qquad Y_2 = \ell_1 C_1 + L_2 C2$$
$$\dot{X}_1 = L_1 C1\dot{\theta}_1 \qquad \dot{Y}_1 = -L_1 S1 \dot{\theta}_1$$
$$\dot{X}_2 = \ell_1 C1 \dot{\theta}_1 + L_2 C2\dot{\theta}_2 \qquad \dot{Y}_2 = -\ell_1 S1 \dot{\theta}_1 - L_2 C_2 \dot{\theta}_2$$

A skeleton of the bond graph can then be constructed and the associated masses, inertias, frictions, elasticities, etc plotted accordingly.

Such a graph physically (but not algebraically) describes a system and, from this, simulations and interpolations can be visualized. In some ways this method is more useful than the method described by Lagrange.

7.4 Difficulties with dynamic models

It should be noted that by using the method described by Lagrange the construction of a dynamic model can be long and tedious and, because the computations are complex, errors are likely. The example quoted in Chapter 3 was a very simple one and if, for example, an end effector possessing three DOF was included into the system, the computations would be considerably more complex (see the example in Chapter 8). This is why some workers have written automatic generation programs of dynamic coefficients in the literal form.

However, a model of a robot is intended not only for simulation purposes but also to enable control systems to be evaluated. In this respect there are three main problems:

1. The model must be sufficiently precise; the more dynamic a model becomes as the speed and acceleration increases, the more important it becomes to consider perturbation effects, such as those arising from:
   (a) errors in length, mass and, above all, moments of inertia.
   (b) elasticity of the different segments and the gearing system (particularly relevant for cable or ribbon-actuated systems).
   (c) backlash and friction, vibration, delays in movement.
   It is difficult to overcome these effects because the model (despite some provision for these effects) may be far from a true representation of the structure of the system (which may change with time, eg the parts may wear). The problem is further complicated by the number of terms in the Lagrange equation.
2. The model must be controllable; however, in practice, some variables cannot be controlled and others require modification of the robot (which can be expensive).
3. In the dynamic model (being essentially non-linear), numerous coefficients of the configurations need to be evaluated in real time. This creates problems when a real-time computer is used because generally they have only a modest speed and capacity. It should be noted that real-time computers are often so slow that their speeds of displacement (which can be controlled in the dynamic mode) often correspond to the validity range of the kinematic model.

It would be useful here to consider how, with the help of a simple example, a dynamic model is developed using equation (7-1), and taking into account one of the sources of oscillation – the elasticity of the cable or belt drive.

7.5 A dynamic model of a belt drive[35]

Consider the transmission between the axis of a motor (pulley of radius r_1 and inertia I_1) and the axis of an articulation (pulley of radius r_2 and inertia I_2) which is moved by a cable with an elasticity of K (see Figure 46).

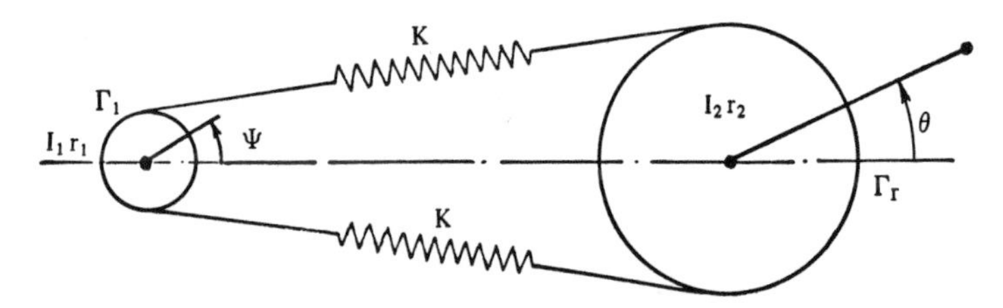

Figure 46. *A simplified representation of transmission by cables*

It has been shown previously[36] that this representation is an acceptable approximation of a more complex belt drive system. The kinetic energy of this system can be written as:

$$2T = I_1 \dot{\Psi}^2 + I_2 \dot{\theta}^2 \tag{7-21}$$

and the internal or potential energy as:

$$U = K(r_1 \Psi - r_2 \theta)^2 \tag{7-22}$$

From the Lagrange factor (L = T − U), equation (7-1) can be written in matrix form as:

$$\begin{pmatrix} I_1 & 0 \\ 0 & I_2 \end{pmatrix}\begin{pmatrix} \ddot{\Psi} \\ \ddot{\theta} \end{pmatrix} + 2K \begin{pmatrix} r_1^2 & -r_1 r_2 \\ -r_1 r_2 & r_2^2 \end{pmatrix}\begin{pmatrix} \Psi \\ \theta \end{pmatrix} = \begin{pmatrix} \Gamma_1 \\ \Gamma_r \end{pmatrix} \tag{7-23}$$

where Γ_1 and Γ_r are the active and resistive torques, respectively. If it is supposed that the resistive torque is zero, the following can be written:

$$a_{11} = -2Kr_1^2/I_1 \quad a_{12} = 2Kr_1 r_2/I_1 \quad u_1 = \Gamma_1/I_1$$
$$a_{21} = 2Kr_1 r_2/I_2 \quad a_{22} = -2Kr_2^2/I_2$$

and two coupled equations can be obtained:

$$\begin{aligned} \ddot{\Psi} &= a_{11}\Psi + a_{12}\theta + u_1 \\ \ddot{\theta} &= a_{21}\Psi + a_{22}\theta \end{aligned} \tag{7-24}$$

The Eigen modes of system (7-24) are given by the equations:

$$(a_{11}-\lambda^2)(a_{22}-\lambda^2)-a_{12}a_{21}=0 \quad (7\text{-}25)$$

$$\lambda^2[\lambda^2-(a_{11}+a_{22})]=0 \quad (7\text{-}26)$$

and the Eigen modes of the oscillation (for the rigid mode) will be $\lambda^2 = 0$, where $\lambda^2 = -2K(r_1^2 + r_2^2)/I_1$ corresponds to an oscillatory mode at the frequency:

$$\Omega = [2K(r_1^2 + r_2^2)/I_1]^{1/2} \quad (7\text{-}27)$$

This mode becomes undesirable if Ω is a mid-range frequency. To control these oscillations the mode must be observable and controllable – it is not enough that equations (7-21) and (7-22) are valid. In the present case its damping requires the addition of sensors to the elastic belt.

Conclusions

It is difficult to produce a dynamic model of a robot, even in the case of rigid, perfectly articulated segments. It is not always possible to use a model to determine the most suitable way of controlling high-speed movements – not only for reasons of validity of the model, but also because of 'real-time' constraints.

Chapter 8

Dynamic control of articulated robots

Dynamic control is concerned mainly with time variables. For example, consider a trajectory $\underline{X}^{(0)}(t)$ in the task space ($\underline{X}$ is the vector defined in Chapter 5); to produce movement along this trajectory a torque vector $\underline{\Gamma}(t)$ must be applied to the actuators or to the articulations (the size of the torque must be known). In practice, movement from $\underline{X}_1$ to $\underline{X}_2$ along trajectory $\underline{X}$ (without oscillation) in a time equal to or less than t_{12} is difficult to achieve.

The computer associated with the system will begin by generating a sample trajectory either in the task space [$\underline{X}(t_i)$ and its time derivatives], or in the articulated variable space [$\underline{\Theta}(t_i)$ and its time derivatives]. This is called 'planning' – it is much more complex than stated previously. It calls for techniques of artificial intelligence because the associated problems can be quite complex and the robot must 'think it over' in order to plan the 'correct' trajectories. At a theoretical level, the planning required for this present (limited) problem involves the same principles as the planning used in setting up a more sophisticated strategy. In the present case, the planner must take into account the following constraints (and ensure that they are adhered to):

1. That $\underline{X}(t)$ is totally contained within the reachable space of the manipulator and that any obstacles are avoided.
2. That $\dot{\underline{\Theta}}(t)$ and $\ddot{\underline{\Theta}}(t)$ involve values of the torques $\underline{\Gamma}(t)$ which are compatible with the attainable maximum values.
3. That, in passing from $\underline{X}$ to $\underline{\Theta}$, the problems of coordinate transformations (and the solutions considered in Chapters 4 and 5) are taken into account.

Real-time control therefore demands the use of a dynamic model at least as complex as the one described by equation (7-11):

$$\underset{N\times N}{[A(\underline{\Theta})]}\,\underset{N\times 1}{\ddot{\underline{\Theta}}} + \underset{N\times C_N^2}{[B(\underline{\Theta})]}\,\underset{N\times 1}{\underline{\dot{\Theta}\dot{\Theta}}} + \underset{N\times N}{[C(\Theta)]}\,.\,\underset{N\times 1}{\dot{\underline{\Theta}}^2} = \underset{N\times 1}{\underline{Q}(\Theta)} + \underset{N\times 1}{\underline{\Gamma}_\theta} \qquad (7\text{-}11)$$

For computer control, the main problem is one associated with computation time. This problem may be overcome by a number of methods.

1. By simplifying the model: This can be done by making hypotheses or statements of the physical significance of some of the values involved and declaring them to be negligible, or by making numerical evaluations and verifying that some terms have no significant variation or are negligible compared with others. For example, it is common practice to neglect the centrifugal and Coriolis forces (although this is not always possible) and to reduce equation (7-11) to the form:

$$[A(\underline{\Theta})]\ddot{\underline{\Theta}} = \underline{Q}(\underline{\Theta}) + \underline{\Gamma}_{\theta} \qquad \textbf{(8-1)}$$

The main disadvantage with this method is that the model becomes less valid.

2. *By calculating some terms in advance:* When the plan, which is stored in the computer, is used to define the successive configurations of the robot it can also be used to evaluate the terms in the matrices $[A(\underline{\Theta})]$, $[B(\underline{\Theta})]$, $[C(\underline{\Theta})]$ and $[Q(\underline{\Theta})]$, and their inverses, as required. It can then store this information and use it during the execution of a task. This has two undesirable consequences:

(a) the planning time increases and, therefore, the operation time.
(b) the stored information can occupy hundreds of kbytes of memory.

The choice of terms to be pre-calculated and the sampling rate should be considered carefully.

3. *By finding a powerful self-correcting control:* The general structure of such a control is illustrated in Figure 47, where the actual and ideal procedures needed to correct movements are compared. There are many alternatives; these are dependent on the variables chosen to characterize the actual evolution and the desired evolution, on the space in which these variables are defined (task space or generalized coordinate space), and on the time of comparison (use of predictive control, for example). In general, self-correcting control improves the

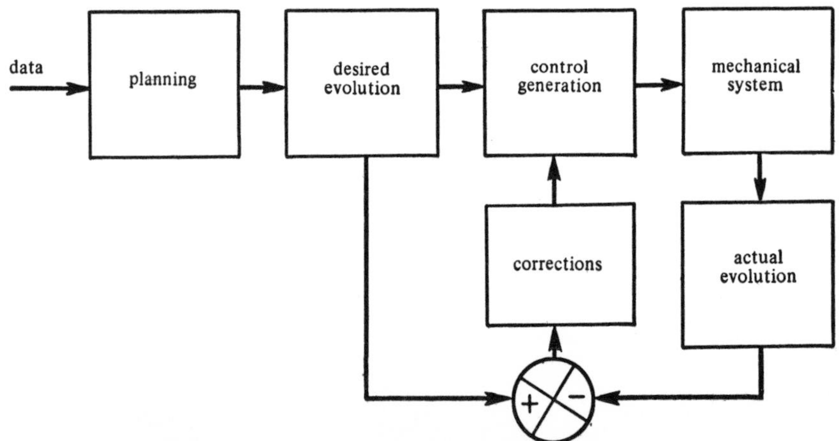

Figure 47. *The principles of self-correcting control*

precision more than it increases speed. However, it can have a positive influence on speed since self-correction may allow the use of a simplified model and, therefore, reduce the computation time.

4. By improving the computer's computation speed: This can be brought about at two complementary levels:

(a) The technological level. Speed performance is improving with each generation of computer. This is true for all types of computer, but particularly for microprocessors which might soon become control computers for even quite 'advanced' robots. However, some functions can be handled by hardware (integrated or wired circuits) instead of having to be programmed. This saves much time and further benefits could be achieved by the use of computer structures better adapted to the problems of control. This implies (i) a distributed computer structure, (ii) a number of processors working in parallel and managed by one supervisor, and (iii) the use of non-sequential but associative processors.

(b) The software level. Computation time can be reduced in several ways: (i) by using a language which is adapted to the problem and to the computer; for example, the use of FORTRAN should be avoided if interested in time performance; (ii) by optimizing the mathematical methods used; for example, in reversing a square matrix several methods may be used – they are all efficient but the one requiring the least number of operations in the computer should be chosen; the sequence of computations used with respect to time criteria should also be optimized; (iii) by optimizing the presentation of formulae [the terms that are numerically evaluated almost always have the form $\Sigma\Pi$, Π, representing a product of constants and trigonometrical values (see, for instance, section 8.2)]; in order to reduce the time of computation, optimal factorization should be obtained; and (iv) by decoupling the equation of the model.

The use of equation (7-11), taking time as the criterion, is made difficult by the coupling which exists betwen the N equations, or in other words, matrices A and C are not diagonal and matrix B cannot be assumed to be zero.

If the equations could be made independent, it would reduce the time of computation since the equations could be treated in parallel, thus simplifying the computations. In fact, non-linear decoupling is possible, as will be shown later.

There are four main ways in which a reduction in the computation time can be achieved (see Table B). First, appropriate software can be used which would allow the use of the original hardware. Alternatively, the mechanical structure can be modified. There are, in fact, a number of ways of achieving a reduction in computation time. However, until now these have never been exploited; most workers have conducted research into the simplification of equations and self-correction rather than into the optimal use of control algorithms.[37-39]

Later in this chapter the difficulties associated with the application of classical dynamic control will be considered and the results obtained by researchers will be discussed.

Mechanism	Structural modification allowing the performance of a pre-determined task while simplifying the model
Model	Simplification of the model Changes to the model
Control algorithm	Changes of algorithm Changes to computation
Changes to computation structures	*Software modifications* – change of language – improvement of the program – change of mathematical method – change of calculation order – change of presentation of the formulae *Hardware modifications* – wired functions – associative computers – division of the programming structure – parallel processors – more rapid computers

Table B. *Principal methods of reducing the time required for control generation using a computer*

8.1 Problems associated with real time and computation

The problems associated with computational delay can be illustrated by consideration of the manipulator shown in Figure 20. Assuming that the hypotheses necessary for planning and a perfected mathematical model can be verified, attempts can be made to make the manipulator follow, in open loop, trajectories specified by position and speed. Any errors in tracking would then be due to delays in computation time and/or from the approximations made in expressing the control algorithm. However, it can be supposed that the level of error derived from the latter is negligible. A classical method of control algorithm examination, based on the state vector of the system, will now be considered.

8.1.1 EVOLUTION OF THE STATE VECTOR

Suppose that the manipulator can be (perfectly) represented by the vectorial mode:

$$\underset{6x6}{[A(\underline{\Theta})]}\ \underset{6x1}{\ddot{\underline{\Theta}}} = \underset{6x1}{\underline{Q}(\underline{\Theta})} + \underset{6x1}{\underline{\Gamma}_\theta} \tag{8-2}$$

Suppose also that the torques ($\Gamma\theta_i$) which act on the articulations are related to the control voltages (V_i) through the transmissions (without backlash, friction, elasticity or mass), and that there is perfect servo control — the following vectorial relationship can then be written:

$$\underset{6x1}{\underline{\Gamma}_\theta} = \underset{6x6}{[C_1]}\underset{6x1}{\underline{V}} + \underset{6x6}{[B_1]}\underset{6x1}{\underline{\Theta}} + \underset{6x6}{[B_2]}\underset{6x1}{\dot{\underline{\Theta}}} \tag{8-3}$$

where C_1, B_1 and B_2 are matrices with constant coefficients. Equations (8-2) and (8-3) provide the equation:

$$\underset{6x6}{[A(\underline{\Theta})]}\underset{6x1}{\ddot{\underline{\Theta}}} = \underset{6x1}{\underline{Q}(\underline{\Theta})} + \underset{6x6}{[C_1]}\underset{6x1}{\underline{V}} + \underset{6x6}{[B_1]}\underset{6x1}{\underline{\Theta}} + \underset{6x6}{[B_2]}\underset{6x1}{\dot{\underline{\Theta}}} \tag{8-4}$$

and the state equation ($x_i = \theta_i$; $x_{i+6} = \dot{x}_i$; $i = 1, 6$) provides:

$$\underset{12x12}{\begin{pmatrix} \Uparrow & \mathbb{O} \\ \hline \mathbb{O} & A \end{pmatrix}}\ \underset{12x1}{\dot{\underline{X}}} = \underset{12x12}{\begin{pmatrix} \mathbb{O} & 1 \\ \hline B_1 & B_2 \end{pmatrix}}\ \underset{12x1}{\underline{X}} + \underset{12x6}{\begin{pmatrix} \mathbb{O} \\ \hline C_1 \end{pmatrix}}\ \underset{6x1}{\underline{V}} + \underset{12x6}{\begin{pmatrix} \mathbb{O} \\ \hline Q \end{pmatrix}} \tag{8-5}$$

where $\underline{X} = (\theta_1, \ldots \theta_6, \dot{\theta}_1, \ldots \dot{\theta}_6)^T$. Consider the simple matrix:

$$[AA]\underline{X} = [BB]\ \underline{X} + [C]\ \underline{V} + [F] \tag{8-6}$$

where A is the matrix of the kinetic energy of the system; equation (8-6) can be inverted to give the vectorial state equation:

$$\dot{\underline{X}} = [AA]^{-1}\ [BB]\underline{X} + [AA]^{-1}\ [C]\underline{V} + [AA]^{-1}\ [F] \tag{8-7}$$

8.1.2 SOLUTION OF THE STATE EQUATION

The theory of automatic control systems[40] allows the solution of equation (8-7) to be written in a continuous form as a function of time:

$$\underline{X}(t) = e^{[AA]^{-1}\ [BB]\ (t-t_0)}\ \underline{X}(t_0) + \int_{\lambda=0}^{t-t_0} e^{[AA]^{-1}\ [BB]\ \lambda}\ [AA]^{-1}\ [C]\ \underline{V}(\lambda)d\lambda$$
$$+ \int_{\mu=0}^{t-t_0} e^{[AA]^{-1}\ [BB]\mu}\ [AA]^{-1}\ [F]d\mu \tag{8-8}$$

Computer control is necessarily discrete, ie a control vector being applied at an instant t, can be changed only at an instant $t + \rho$, ρ being the sample period. Quantification of equation (8-8) gives the vectorial equation:[41]

$$\underset{12x1}{\underline{X}_{(K+1)T}} = \underset{12x12}{[G_K(T)]}\underset{12x1}{\underline{X}_{KT}} + \underset{12x6}{[R_K(T)]}\underset{6x1}{\underline{V}_{KT}} + \underset{12x6}{[H_K(T)]} \quad \textbf{(8-9)}$$

with

$$G_K(T) = e^{[AA]_K^{-1}\,[BB]_K\,T} \quad \textbf{(8-10)}$$

$$R_K(T) = \int_0^T e^{[AA]_K^{-1}\,[BB]_K\,\lambda}\,[AA]_K^{-1}\,[C]_K\,d\lambda \quad \textbf{(8-11)}$$

$$H_K(T) = \int_0^T e^{[AA]_K^{-1}\,[BB]_K\,\mu}\,[AA]_K^{-1}\,[F]_K\,d\mu \quad \textbf{(8-12)}$$

where KT are the times of interest. These expressions can be evaluated from limited developments of e^x, eg up to the second or third order.

8.1.3 THE CONTROL ALGORITHM

It should be noted that in equation (8-9) the state vector ($\underline{X}$) has 12 components, whereas the control vector ($\underline{V}$) has only six. The theory governing the control of the system assumes that to make a system pass from one state (X_0) to another (X_1), the lowest number of control samples will be 12/6 = 2. Equation (8-9) is in state $\underline{X}$ and can be used iteratively on two successive samples to give equation (8-13); by changing the notation KT to K, this can be simplified to:

$$\underset{12x1}{\underline{X}_{K+2}} = \underset{12x12}{[G_{K+1}\,G_K]}\underset{12x1}{\underline{X}_K} + \underset{12x12}{[G_{K+1}\,R_K,\,R_{K+1}]}\underset{12x1}{\begin{pmatrix}\underline{V}_K\\ \underline{V}_{K+1}\end{pmatrix}} + \underset{12x6}{[G_{K+1}\,H_K + H_{K+1}]} \quad \textbf{(8-13)}$$

To pass from state $\underline{X}_K$ to state $\underline{X}_{K+2}$ a control vector needs to be applied which can be defined by:

$$\begin{pmatrix}\underline{V}_K\\ \underline{V}_{K+1}\end{pmatrix} = [G_{K+1}\,R_K,\,R_{K+1}]^{-1}\,[\underline{X}_{K+2} - (G_{K+1}\,G_K)\,\underline{X}_K - (G_{K+1}\,H_K + H_{K+1})] \quad \textbf{(8-14)}$$

this assumes that $[G_{K+1}\,R_K,\,R_{K+1}]$ allows inversion. Equation (8-14) constitutes a mathematical control algorithm.

8.1.4 PROBLEMS ASSOCIATED WITH REAL-TIME CONTROL

Suppose that the task can be described in the space of articulated variables in a continuous form with respect to time: $X(t) = [\underline{\Theta}(t), \dot{\underline{\Theta}}(t)]$.

Equation (8-14) guarantees transition only through the points $X(k'T) = [\dot{\underline{\Theta}}(k'T), \underline{\Theta}(k'T)]$. The smaller the sample period (T), the better the tracking. In practice the sample period is expressed in milliseconds. It can be supposed that the trajectory $\underline{X}(t)$ is correctly sampled, ie at intervals in time and space which are compatible with the robot's electromechanical capabilities, and with the necessary precision – $\underline{X}(t)$ being compatible with these requirements.

8.1.4.1 Open-loop control

Once that trajectory sampling has been carried out then all the elements comprising the right-hand side of equation (8-14) are known. Therefore, it is easy to calculate in advance the successive control vectors. In principle, it is bound to pass exactly through one of the two points ($\underline{X}_K$, $\underline{X}_{K+2}$, etc). Under such conditions, all the successive control vectors being evaluated off-line, the computation time will not be too great since, on-line, only stored successive values of $\underline{V}_K$ need to be retrieved from the computer memory. However, if equation (8-4) is imperfect, there is a possibility of divergence between the actual and required trajectory – this cannot be corrected and the divergence increases with time.

8.1.4.2 Correction by determination of the real value of $\underline{X}_K$

If all the elements of equation (8-14), except those which involve $\underline{X}_K$, are evaluated off-line and then stored (which considerably increases the amount of memory occupied compared with the previous case), there will be 144 multiplications and 144 additions performed on-line. Using a minicomputer, the time taken to perform these simple operations would be in the range of 9 (microprogrammed multiplication) to 2 ms (wired multiplication). However, in the present case this time could be halved [noting that equation (8-14) can be fixed] by first evaluating $\underline{V}_K$. $\underline{X}_K$ should be measured first and then $\underline{V}_K$ (computed from $\underline{X}_K$) applied to obtain a correction factor for equation (8-14).

8.1.4.3 Other corrections

Equation (8-14) does not constitute a perfect model. The approximation considered in equations (8-9) to (8-12) risk cumulative errors. However, the method described in the previous section requires the use of a computer with a considerable memory size. A better result would be obtained if, knowing $\underline{X}_{K+2}$ and measuring $\underline{X}_K$, at that instant, all the other elements in equation (8-14) could be evaluated. However, it should be noted that the computation time exceeds 100 ms which is unacceptable (this is common when a minicomputer is used). This could be solved by changing algorithm (8-14) so that a delay occurs between the measurement of $\underline{X}_K$ and the application of $\underline{V}_K$.

8.1.5 MODIFICATION OF THE CONTROL EQUATION

The following modification is a good example: if the interval (K, K + 2) is divided into four parts, K to K + 4, and $\underline{V}_K = \underline{V}_{K+2} = 0$ is imposed, then a new algorithm can be obtained:

$$[\underline{V}_{K+1}, \underline{V}_{K+3}]^T = [G_{K+2}(2T')\,R_K(T'),\, R_{K+2}(T')]^{-1}\,[\underline{X}_{K+4} - G_{K+2}(2T')\,G_K(2T')\underline{X}_K$$

$$- G_{K+2}(2T')\,H_K(2T') - H_{K+2}(2T')] \qquad \textbf{(8-15)}$$

where $\underline{V}_K = \underline{V}_{K+2} = 0$, and T is the new period (= T/2). The algorithm can then be produced according to the scheme shown in Figure 48.

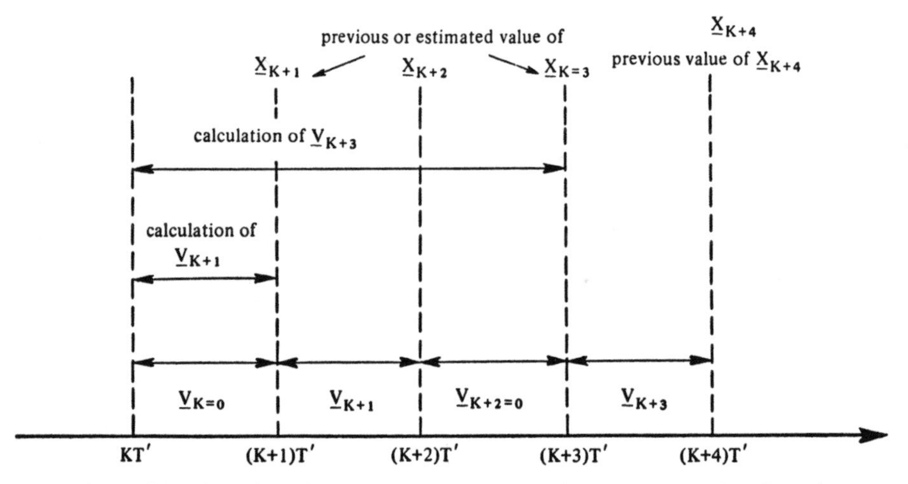

Figure 48. *The principles of algorithmic execution using equation (8-15)*

8.1.6 EXPECTED VALUES FOR THE PERFORMANCES

From the instant kT', $\underline{V}_{K+1}$ would have to be evaluated in time T'. It is this computation which limits the quality of dynamic tracking. All the elements relevant to the time taken are presented in Table B. The following average values have been assumed: time (in microseconds) taken for addition = 3, multiplication = 7, division = 8, and computation of trigonometrical functions = 10.

Using the Gauss algorithm for matrix inversion (without changing the organization of the computation) a computation delay of $\underline{V}_{K+1}$ can be derived from equation (8-15) [all the elements other than $\underline{X}_K$ (measured) and $\underline{X}_{K+4}$ (required) are calculated on-line] — this delay is usually about 55 ms. Robot sequences can, at most, be separated by time intervals of 220 ms. Consequently, it should be possible to ensure positional control of a robot's terminal position only at one point every 20 cm, at a displacement velocity of 1 m/s. When it is necessary to control a trajectory (by positional control) every 2 cm, it is not possible to exceed 10 cm/s – for which, in practice, kinematic control is sufficient. The time of computation can be reduced by evaluating in advance some of the terms of equation (8-15).

For instance, if A_K, A_{K+2}, $\underline{Q}_K$ and Q_{K+2} are evaluated in advance, the time period (T_0) declines to about 160 ms (one point every 16 cm at 1 m/s), ie the improvement is slight. However, if all the terms (except $\underline{X}_K$) in equation (8-15) are known (and it continues to increase as described in section 8.1.4.2), then T_0 declines to 30 ms (one point every 3 cm at 1 m/s), but the size of the memory needed for storage of the terms evaluated in advance becomes considerable and the risk of error increases.

8.1.7 COMPARISON OF THE PRECISION OF DYNAMIC AND KINEMATIC TRACKING MODES

In the dynamic mode, the instantaneous speed and the position are controlled, whereas in the kinematic mode the position and, indirectly, the average speed are controlled.

It is interesting to compare the positional precision of a dynamic algorithm with that of a kinematic algorithm when the period of sampling (the speed) is varied. There are detailed examples of such comparisons in the literature;[42] a few of the results obtained will be noted here.

The specific trajectories (defined in the articulated variable space for the manipulator shown in Figure 20) are first spatially sampled. Then each of the sample points are evaluated at different average speeds by varying the period T_T – the time taken to move from one point to the next.

All the computations are made in advance when producing the algorithm of dynamic tracking; therefore, successive values of $\underline{V}_K$ are known. The algorithm of kinematic tracking is produced by considering the static equilibrium at each successive point of the trajectory. Equation (8-4) can be reduced to:

$$\underline{V} = -[C_1]^{-1}[B_1]\underline{\Theta} - [C_1]^{-1}\underline{Q}(\underline{\Theta}) \qquad (8\text{-}16)$$

$$\underline{V}_K = -[C_1]^{-1}[B_1]\underline{\Theta}_K - [C_1]^{-1}\underline{Q}_K(\underline{\Theta}_K) \qquad (8\text{-}17)$$

Positional error is defined as the quadratic mean error of all the reference points of a trajectory, and all the variables (θ_i):

$$\overline{E(\underline{\Theta})} = \sum_{i=1}^{6} \frac{\left[\sum_{n=1}^{R} (\theta_{in}^{d} - \theta_{in}^{r})^2\right]^{1/2}}{R} \qquad (8\text{-}18)$$

where R is the number of points on the trajectory, θ_{in}^{d} the target value for the variable at n, a particular point on the trajectory, and θ_{in}^{r} the value of θ_i at point n. Simulation gives results of the form illustrated in Figure 49.

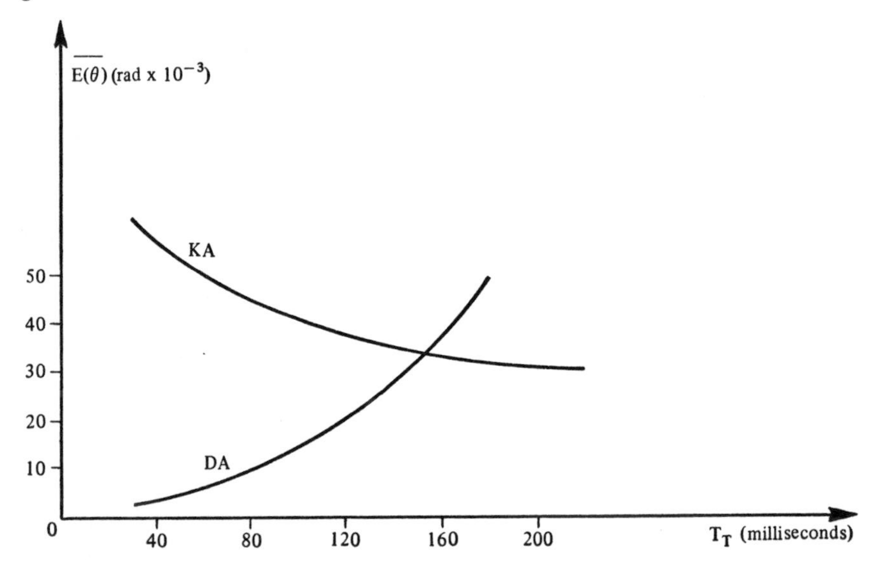

Figure 49. *A comparison of the precision obtaining in positional tracking of a trajectory using a kinetic algorithm and a dynamic algorithm as a function of the sampling period (T_T) – the average varied as much as $1/T_T$*

It is apparent (see Figure 49) that the dynamic algorithm performs poorly at high values of T_T for small sampling periods (at high speeds) whereas the reverse is true for the kinematic algorithm. This can be explained by:

(a) for the kinematic algorithm, the lower the average speed, the more valid the hypotheses of static equilibrium at the points under consideration become.

(b) for the dynamic algorithm, the greater the distance between two successive states, the less valid the approximations used (limited development of exponentials, invariance of the dynamic coefficient, etc) become. In practice, computation delay sets a minimum value for T_T. If this value is too great, the dynamic algorithm is no longer valid.

8.1.8 CONCLUSIONS

This example has indicated the limitations which result from computational delays; advanced control systems for a robot using dynamic intelligence (interaction with the environment, pattern recognition, artificial intelligence, etc) should be designed with this phenomenon in mind. Good control must be achieved as quickly as possible and any computational delays minimized. These two criteria are not independent however, and improvements in the quality of control generally increase computation times. Optimization is a complex matter, the number of parameters involved being large.

8.2 Simplification of the equation of the model

The time taken for a computer to compute the equation of the model can be reduced by minimizing the number of computations involved (by simplifying the model).

Consider the structure illustrated in Figure 20. The elements of this structure can be represented by the system shown in Figure 50 (which shows a type MA 23 machine, Calhent-CEA).[43] The robot possesses six rotational DOF. The matrices A, B and C of equation (7-11) contain 36 terms a_{ij}, 90 terms b_{ijk} and 36 terms c_{ij}, respectively, a total of 162 complex functions of θ_i.

To obtain some idea of this complexity (which is common to all robots possessing six DOF) the following two coefficients should be studied – one being the least complex, not equal to zero (c_{54}), and the other very complex (a_{11}):

$$c_{54} = M_5(-C5.\ell_{Y_4}.d_5 - SC5.d_5^2 + SC5.I_{Z_5}) \tag{8-19}$$

$$\begin{aligned}
a_{11} = {} & I_{Z_1} + M_2\,SS2\,d_2^2 + SS2.I_{Y_2} + CC2.I_{Z_2} + M_3\,[SS2\,\ell_2^2 \\
& + 2S2S\,(2{+}3)\,C_2 d_3 + SS\,(2{+}3) d_3] + SS(2{+}3)\,I_{Y_3} \\
& + CC(2{+}3)\,I_{Z_3} + M_4\,[SS2\,\ell_2^2 + 2S2S(2{+}3)\,C_2\,h_1 \\
& + 2S2C(2{+}3)\,C4\,C_2 d_{Y_4} + SS(2{+}3) h_1^2 + 2SC(2{+}3)\,C4\,d_{Y_4} h_1 \\
& + CC(2{+}3)\,d_{Y_4}^2 + SS\,(2{+}3)\,SS4 d_{Y_4}^2] + SS(2{+}3)\,SS4\,I_{X_4} \\
& + SS(2{+}3)\,CC4.I_{Y_4} - 2SC(2{+}3)\,C4\,I_{YZ_4} + CC(2{+}3)\,I_{Z_4} \\
& + M_5\,[SS2\,\ell_2^2 + 2S2S(2{+}3)\,C_2 h_2 + 2S2C\,(2{+}3)\,C4\,C_2\,C_{Y_4} \\
& + 2\,S2\,S(2{+}3)\,C5\,C_2 d_5 + 2S2C(2{+}3)\,C4S5\,C_2 d_5 + SS(2{+}3) h_2^2 \\
& + 2SC(2{+}3)\,C4\,C_{Y_4} h_2 + 2SS(2{+}3)\,C5\,h_2 d_5 + 2SC(2{+}3)\,C4S5\,h_2 d_5 \\
& + CC(2{+}3)\,\ell_{Y_4}^3 + 2SC(2{+}3) C4C5\,\ell_{Y_4} d_5 + 2CC(2{+}3)\,S5\,\ell_{Y_4} d_5 \\
& + SS(2{+}3) CC4CC5\,d_5 + 2SC(2{+}3)\,C4SC5\,d_5^2 + CC(2{+}3)\,SS5\,d_5^2] \\
& + SS(2{+}3)\,SS4\,C_{Y_4}^2 + 2SS\,(2{+}3)\,SS4S5\,\ell_{Y_4} d_5 + SS(2{+}3)\,SS4\,d_5^2 \\
& + SS(2{+}3)\,SS4\,I_{X_5} + SS(2{+}3)\,CC4CC5\,I_{Y_5} + 2SC(2{+}3)\,C4SC5\,I_{Y_5} \\
& + CC(2{+}3)\,SS5\,I_{Y_5} + SS(2{+}3)\,CC4SS5\,I_{Z_5} - 2SC(2{+}3)\,C4SC5\,I_{Z_5} \\
& + CC(2{+}3)\,CC5\,I_{Z_5} + M_6\,[SS2\,\ell_g^2 + 2S3S\,(2{+}3)\,\ell_g d_6 \\
& + SS(2{+}3) d_6^2] + SS(2{+}3)\,I_{Y_6} + CC(2{+}3)\,I_{Z_6}
\end{aligned} \tag{8-20}$$

The notations used are: SI ≡ sin θ_i; CI ≡ cos θ_i; SCI ≡ sin θ_i cos θ_i; SSI ≡ $\sin^2 \theta_i$; CCI ≡ $\cos^2 \theta_i$; M_i is the mass of the lever (i); $I_{X\lambda}$, $I_{Z\lambda}$ are the inertias of the segments (λ) about the axes X, Y and Z, respectively, which are bound to this segment; and $I_{YZ\lambda}$ is the inertia of the segment (λ) with respect to plane YX.

An appropriate form of equation (7-12) can be written for the manipulator:

$$\begin{cases} \sum_{j=1}^{6} [A(i,j)\,\ddot{\theta}_j + C(i,j)\,.\,\dot{\theta}_j^2 + \sum_{k=j+1}^{6} B(i,j,k)\,\dot{\theta}_j\,\dot{\theta}_k] = Q(i) + \Gamma_{\theta_i} \\ i = 1,6 \end{cases} \tag{8-21}$$

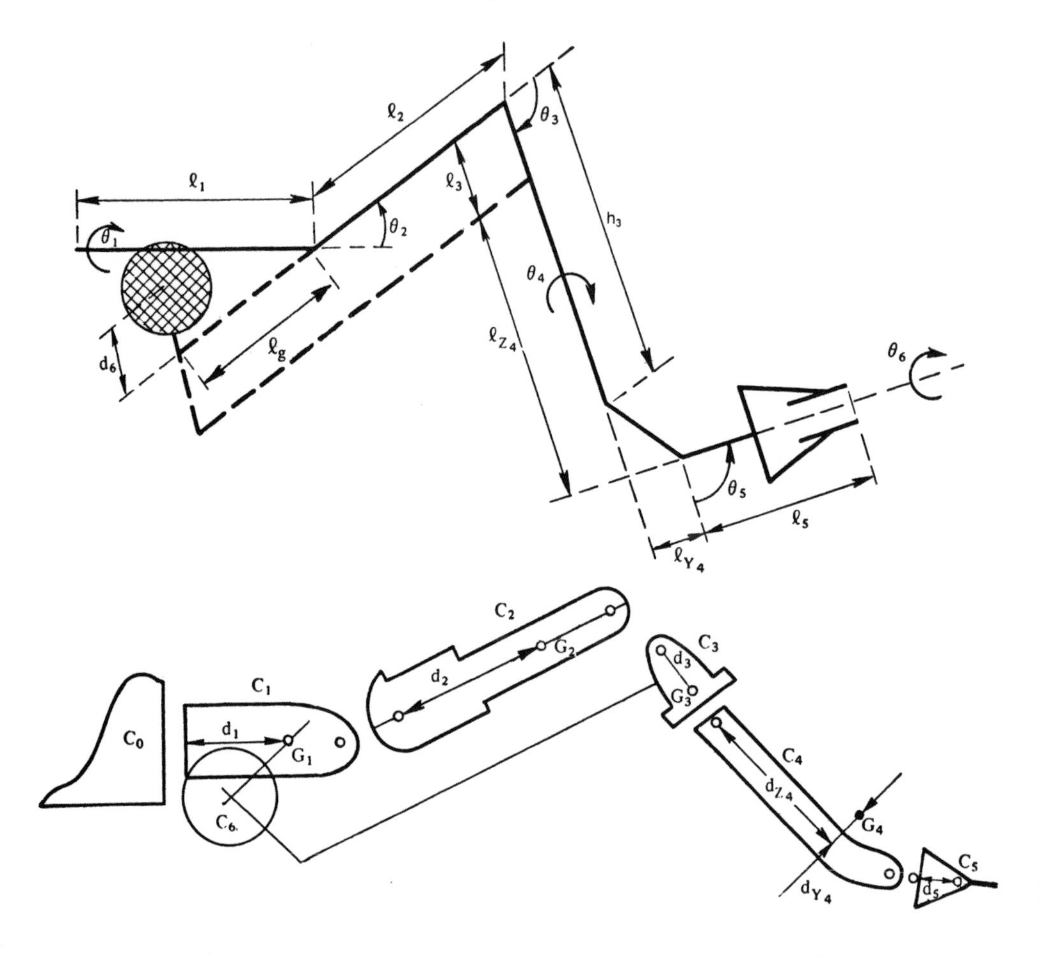

Figure 50. *The parameters which are included in the terms of the dynamic coefficients in matrices A, B and C*

This system can be simplified by determining whether some of the terms are equal to zero, equal to one another, or negligible. There are three methods[44] available:

1. The general machine theory. This assumes that:

(a) the coefficient A(i, j) of the reaction torque generated by an acceleration of θ_j on the movement of θ_i is equal to the coefficient A(j, i) of the reaction torque generated by the acceleration of θ_i on the movement of θ_j; thus, A(i, j) = A(j, i) and the matrix [A] is symmetrical.

(b) the centrifugal force has no effect on the movement of the articulation that generates it, ie C(i, j) = 0.

(c) the torques resulting from the interaction between the speed of θ_i and the speeds of segments situated above θ_i are zero;

this assumes that B (i, k, i) = 0 if i > k – this is taken into account in the formulation of equation (8-21).

The first assumption, which can be applied to all mechanisms, allows a reduction in the number of dynamic coefficients to be calculated – in the present case, from 162 to 126.

2. *Consideration of the mechanical design.* For the present example:

(a) consider the unloaded manipulator (or if the load carried by the gripper is relatively small, with its centre of gravity situated at the gripper's rotational axis corresponding to θ_6), all the coefficients that imply that θ_6 is coupled with another variable θ_i, are zero. A(i, 6), C(i, 6), C(6, i), B(i, j, 6), B(6, j, k) are invalid; i, j, k are different from 6, ie 51 terms. Similarly, A (4,5) = C (4,5) = 0.

(b) the parallelism of the rotational axes of θ_2 and θ_3 leads to the invalidity of certain coefficients:

$$B_{213} = 0 \qquad B_{i24} = B_{i34}$$
$$B_{i23} = 2C_{i3} \qquad B_{i25} = B_{i35}$$
$$B_{412} = -B_{214} \qquad B_{413} = -B_{314}$$
$$B_{512} = -B_{215} \qquad B_{513} = -B_{315}$$

On the whole, a particular mechanical design of the robot can produce an economy of 70 terms. At this stage of the computations there are 56 coefficients compared with the initial number of 162.

3. *Consideration of maximum speeds.* The speed of movement is limited for reasons of technology or by constraints imposed by the operator. Using simulation techniques many of the coefficients can be invalidated. In the present case, if the load is no more than 50 N and the speed of the end effector no more than 1 m/s, then the centrifugal and Coriolis forces become negligible. Then only 15 terms A(i, j) and six terms Q(i) need to be calculated. The process is reviewed in Table C.

Under such conditions, a valid model can be written as:

$$\sum_{j=1}^{6} A(i,j)\,\ddot{\theta}_j = Q(i) + \Gamma_{\theta_i} \qquad i = 1,6 \tag{8-22}$$

However, in industrial applications, movements are executed as quickly as possible and the third assumption is not always valid. Fournier[45] quoted an example in which a simulated robot (see Figure 7) with ten articulations and possessing six DOF was examined.

For a given trajectory (maximum speed of point O_g is about 1.5 m/s, acceleration is about 15 m/s), the evolution of the torques developed by the first three articulations can be described by classifying their origins as being due to acceleration, centrifugal force or Coriolis force.

Dynamic coefficients	Initial number	Decrease in the number to be calculated		Final number
		due to	*number*	
A (i, j)	$n^2 = 36$	general theorems	$\frac{n(n-1)}{2} = 15$	21
		mechanical design	6	15
		maximum speed	0	15
B (i, j, k)	$\frac{n^2(n-1)}{2} = 90$	general theorems	$\frac{n(n-1)}{2} = 15$	75
		mechanical design	decoupling θ_6 : 35 parallel axes : 11	40 29
		maximum speed	29	0
C (i, j)	$n^2 = 36$	general theorems	$n = 6$	30
		mechanical design	15	15
		maximum speed	15	0
Total	$\frac{n^3 + 3n^2}{2} = 162$		147	15

Table C. *Example of the possible reduction of the number of dynamic coefficients to be calculated for control*

Figures 51 to 53 represent the three torques to which the second articulation is subjected during tracking. In Figures 54 and 55 the relation between the centrifugal and Coriolis torques with the inertial couple has been illustrated. It should be noted that, depending on the time factor, these relationships can be significant, having ratios > 1. Under such conditions, it is not possible to simplify the model by suppressing terms B(i, j, k) and C(i, j).

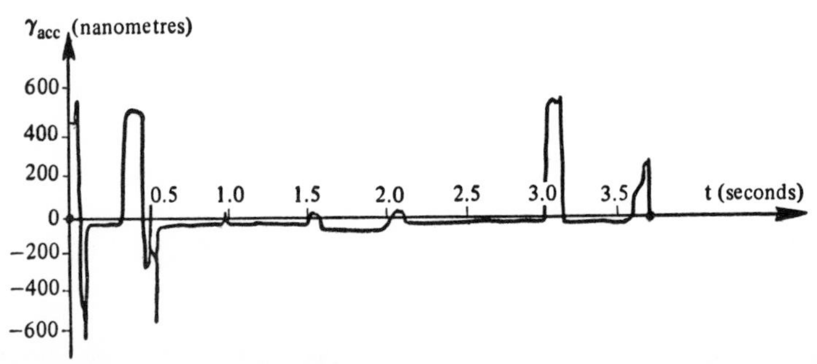

Figure 51. *Evolution of the tracking torque on articulation 2 during tracking of a trajectory defined by the following (successive) points expressed in R_0 (in metres): (X = 0; Y = −1.3; Z = 2.15), (−0.25; −1; 1.80), (−0.2; 2; 1.80), (0.25; −1.95; 1.80), (0.2; −1; 1.80) and (−0.3; −1; 1.80)*

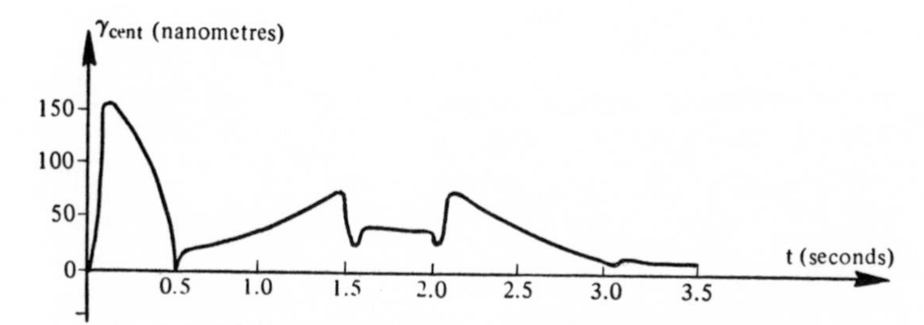

Figure 52. *Evolution of the torque due to centrifugal forces in the articulation during execution of the trajectory*

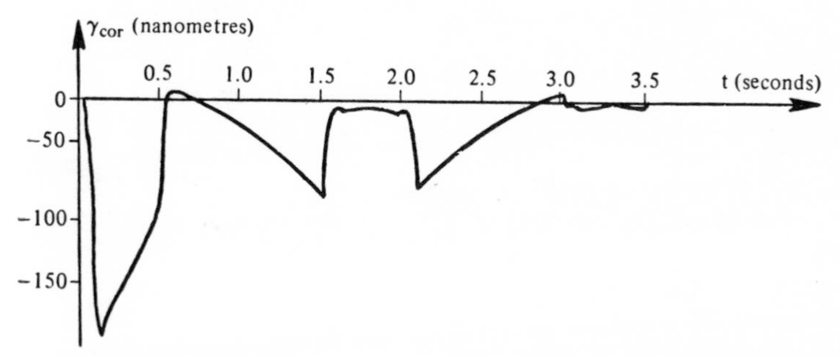

Figure 53. *Evolution of the torque due to coupling forces on the articulation during execution of the trajectory*

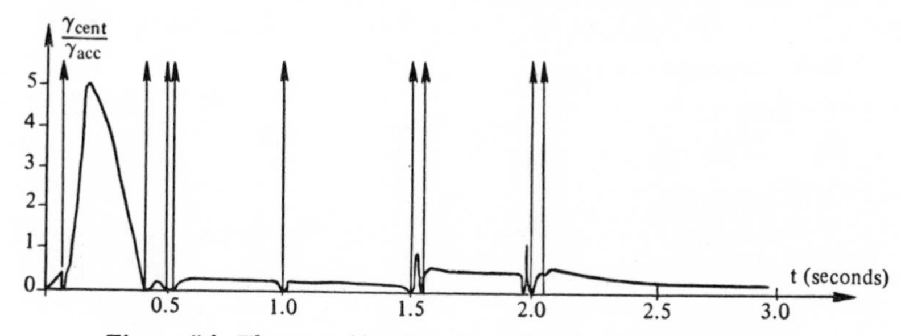

Figure 54. *The centrifugal and acceleration torque ratio against time during tracking*

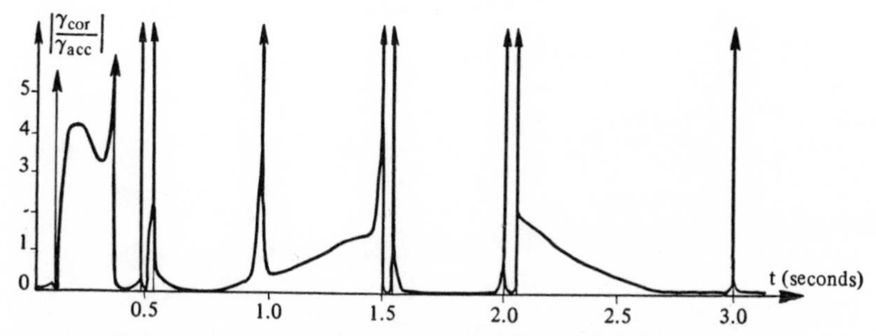

Figure 55. *The relation between* $|\gamma_{cor}/\gamma_{acc}|$ *during the execution of the trajectory*

8.3 Other methods of dynamic control

Modelling does not truly reflect a real situation and computations (some being lengthy) can cause delays. For these reasons, self-correcting methods have been developed which make use of simplified models.

8.3.1 KAHN'S MODEL[46]

In 1970, Kahn described a system of dynamic control of an articulated robot using well-known techniques. It consisted of linearization around an *ad hoc* operating point and 'bang-bang' control. In practice it proved unworkable as linearization cannot be justified when robots are used for a variety of tasks.

8.3.2 PAUL'S METHOD[47]

This system of control takes into account only the coefficients (A_{ii}) of the main diagonal of the matrix A. Thus, the equations are partially decoupled:

$$A_{ii}(\underline{\Theta}) \cdot \ddot{\underline{\Theta}}_i = \underline{Q}(\underline{\Theta}) + \underline{\Gamma}_{\theta_i} \quad \textbf{(8-23)}$$

Paul proposed a supplementary approximation which consists of dividing the trajectory into segments. The values of A_{ii} can then be calculated precisely at the end of each segment and the intermediate points calculated by interpolation. However, this method is valid only if the inertial couplings are negligible – this is so only at low speeds when kinematic control is possible.

8.3.3 BEJCZY'S METHOD[48]

The principles of this method are shown in Figure 56. The velocity and positional corrections (with constant gain) ensure stability in the system. The mathematical model elaborates on the torques at each of the articulations. The performance of the model varies a great deal because linear corrections are imposed on a non-linear system.

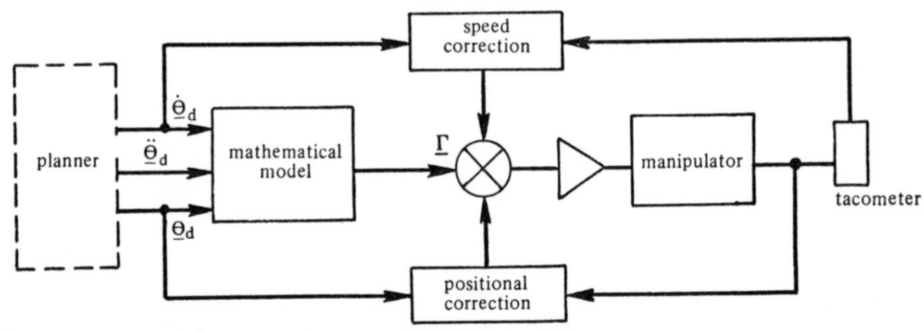

Figure 56. *A scheme of Bejczy's method of dynamic control*

8.3.4 THE SPATIAL CONFIGURATION METHOD[49]

Since there are considerable difficulties in calculating dynamic coefficients on-line, they may be calculated in advance for all possible configurations. In practice the size of the memory required for such an operation is huge. Raibert and Horn suggested that the task space could be divided into zones and that the coefficients could be considered to be constant. This allows the solution of part of the problem as the number of mathematical operations would be reduced to $N^3 + 3N^2$ (where N is the number of DOF). For a system possessing six DOF a memory capacity of 250 kbytes would be required.

8.3.5 KHATIB AND LE MAITRE'S METHOD[50]

This method supposes that the robot moves in a potential field within the task space. It is assumed that the end effector is attracted to its working position and is repelled by any obstacles. The environment can be described as a potential function which adds an artificial force vector to the generalized forces.[51–53]

8.3.6 THE NON-LINEAR DECOUPLING METHOD

Freund showed that the control of any system put in the form:

$$\begin{aligned} \dot{\underline{x}}(t) &= A(x, t) + B(x, t)\,\underline{u}(t) \\ \dot{\underline{y}}(t) &= C(x, t) + D(x, t)\,\underline{u}(t) \end{aligned} \tag{8-24}$$

can be linearized (where A, B, C and D are matrices with compatible dimensions and in which the elements are non-linear functions of the state of the system). Zaballa[54] has shown that every AMS can be modelled in this form.

Khalil and Liegeois[55] obtained linearization of control by using the general equation:

$$A\underline{\ddot{\Theta}} + B\underline{\dot{\Theta}\dot{\Theta}} + C\underline{\dot{\Theta}}^2 = \underline{Q} + \underline{\Gamma} \tag{7-11}$$

To determine a control which can decouple a system, it can be said that:

$$\underline{\Gamma} = \underline{\Gamma}_1 + \underline{\Gamma}_2 \tag{8-25}$$

If compensation for the effect of perturbational terms is attributed to Γ_1, ie torques caused by gravitational, Coriolis and centrifugal forces:

$$\underline{\Gamma}_1 = B\,\underline{\dot{\Theta}\dot{\Theta}} + C\underline{\dot{\Theta}}^2 - \underline{Q} \tag{8-26}$$

then:

$$\underline{\Gamma}_2 = A\,\underline{\ddot{\Theta}} \tag{8-27}$$

To obtain a system consisting of N decoupled, second-order linear equations a value of Γ_2 should be chosen so that:

$$\underline{\Gamma}_2 = A\,[-K_p\underline{\Theta} - K_v\dot{\underline{\Theta}} + \lambda\,\underline{W}(t)] \quad \textbf{(8-28)}$$

where K_p, K_v and λ are diagonal matrices with constant arbitrary coefficients. The use of equations (8-27) and (8-28) then leads to:

$$\begin{cases} \ddot{\theta}_i + K_{pi}\,\theta_i + K_{vi}\,\dot{\theta}_i = \lambda_i\,W_i(t) \\ i = 1, N \end{cases} \quad \textbf{(8-29)}$$

The principles of this type of control are represented by Figure 57.

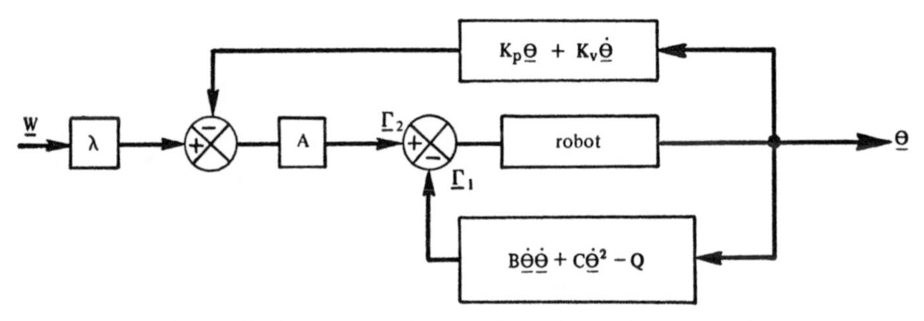

Figure 57. *Control by the non-linear decoupling method*

This method requires the use of a perfect model, otherwise $\underline{\Gamma}_1$ remains unrealizable and decoupling is not achieved.

8.3.7 PREDICTIVE CONTROL[56]

In this method, torques (obtained from known values of the required control variables, and by using the mathematical model and correction values) are applied to the robot. The principles involved are represented by Figure 58.

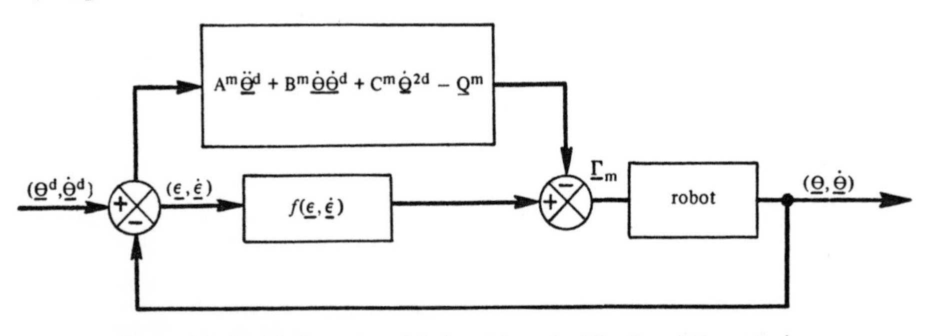

Figure 58. *Predictive control (after Liegeois, Khalil and Fournier)*

The value of $\underline{\Gamma}_m$ can be calculated from the variables $\underline{\Theta}^d$, $\dot{\underline{\Theta}}^d$ and $\ddot{\underline{\Theta}}^d$:

$$\begin{aligned} \underline{\Gamma}_m &= A^m\ddot{\underline{\Theta}}^d + B^m\dot{\underline{\Theta}}\dot{\underline{\Theta}}^d + C^m\dot{\underline{\Theta}}^{2d} - \underline{Q}^m \\ \underline{\Gamma} &= \underline{\Gamma}^m + \underline{f}(\underline{\epsilon}, \dot{\underline{\epsilon}}) \\ \underline{\epsilon} &= \underline{\Theta}^d - \underline{\Theta};\ \dot{\epsilon} = \dot{\underline{\Theta}}^d - \underline{\Theta} \end{aligned} \quad (8\text{-}30)$$

The system can be described by the equation:

$$A\ddot{\underline{\Theta}} + B\dot{\underline{\Theta}}\dot{\underline{\Theta}} + C\dot{\underline{\Theta}}^2 - \underline{Q} = A^m\ddot{\underline{\Theta}}^d + B^m\dot{\underline{\Theta}}\dot{\underline{\Theta}}^d + C^m\dot{\underline{\Theta}}^{2d} - \underline{Q}^m + \underline{f}(\underline{\epsilon}, \dot{\underline{\epsilon}}) \quad (8\text{-}31)$$

The problem which then remains is to determine the ideal function for $\underline{f}$ $(\underline{\epsilon}, \dot{\underline{\epsilon}})$. Suppose that there are small variations in structure between the process and the model, then a first-order development of the dynamic coefficients as a function of these variations gives the equation:

$$A^m\ddot{\underline{\epsilon}} + \underline{f}(\underline{\epsilon}, \dot{\underline{\epsilon}}) = \underline{0} \quad (8\text{-}32)$$

if the variations are invalidated. To obtain linear equations with constant and arbitrary coefficients it is sufficient to write (as in section 8.3.6):

$$\underline{f}(\underline{\epsilon}, \dot{\underline{\epsilon}}) = A^m K_v \dot{\underline{\epsilon}} + A^m K_p \underline{\epsilon} \quad (8\text{-}33)$$

These coefficients are chosen to ensure good translational behaviour and stability when $\underline{\epsilon} = 0$. If the structural variations are not zero, equation (8-32) becomes:

$$A^m \ddot{\epsilon} + \underline{f}(\epsilon, \dot{\epsilon}) = \underline{g}\,[\underline{\Theta}_d\,(t)] \quad (8\text{-}34)$$

Looking for a solution of a type like equation (8-33):

$$\ddot{\underline{\epsilon}} + K_v\dot{\underline{\epsilon}} + K_p\underline{\epsilon} = \underline{h}\,(\underline{\Theta}^d) \quad (8\text{-}35)$$

where $\underline{h}$ can be considered to be a perturbation of the mechanical system on the main variable, $\underline{\epsilon}$. The system gives a permanent response if $\underline{h}$ remains limited and if the energy dissipated by the system $\dot{\underline{\epsilon}}^T K_V \dot{\underline{\epsilon}}$ is significant enough.

This method is of interest because it allows for errors in modelling. The control stays precise even when centrifugal and Coriolis forces are not taken into account by the model (providing, that is, they do not exceed 'reasonable' values).

8.4 The choice of the space used in computation

It has been shown in previous chapters how, to control a robot, two spaces – that of the generalized variable ($\underline{\Theta}$) and that of the task space ($\underline{X}$) – need to be considered. The problems associated with coordinate transformation (passing from one space to another) have also been described. All the methods devised to date (apart from the method of Khatib and Le Maitre) call upon the use of generalized variable space.

Recalling the equations:

$$\underline{X}(R_0) = \underline{F}(\underline{\Theta}-\underline{\Theta}_0) \quad \textbf{(5-1)}$$

$$\dot{\underline{X}}(R_0) = [J]\,\dot{\underline{\Theta}} \quad \textbf{(8-36)}$$

equation (8-36) can be derived from equation (6-5) and its inverse:

$$\dot{\underline{\Theta}} = [J]^{-1}\,\dot{\underline{X}}(R_0) \quad \textbf{(8-37)}$$

and equation (8-37) from equation (6-16). However, this creates the problems discussed in section 6.2. But the following can be written:

$$\ddot{\underline{\Theta}} = [J]^{-1}\,\ddot{\underline{X}}(R_0) - [J]^{-1}\,[\dot{J}]\,[J]^{-1}\,\dot{\underline{X}} \quad \textbf{(8-38)}$$

with:

$$[\dot{J}] = \frac{d}{dt}\,[J] \quad \textbf{(8-39)}$$

Every equation in $\underline{\Theta}$ can, therefore, be transposed in the task space. In some cases, because of the complexity of the equations and the time taken to solve them, the task space can be used.

Conclusions

A knowledge of the dynamic control of robots is essential before they can be successfully integrated into production lines operating at high speeds. Freund, Khatib and Khalil have discussed control at the theoretical level. However, how dynamic control is achieved in practice (using parallel computers, specialized operators and computers with larger memories)[57] is complex.

Chapter 9
Learning and trajectory generation

Here a computer-controlled robot will be considered which is equipped with internal positional sensors and servo control only. Whether kinematic or dynamic control is used it will function as shown in Figure 47 (the planning element has been excluded).

If the robot is to perform a task (t_x) it is necessary to describe this task without entering into task analysis (see Chapter 10).

The operator should:

1. Establish that the task to be performed lies within the space capable of being reached by the robot.
2. Describe the problem both in a natural language and in a condensed form, eg 'shift object A in direction X' and 'lift and place object A in zone B whilst avoiding obstacle C'.

When robots without external, absolute sensors are used, one of two principal forms of operation applies:

1. The operator controls the robot manually and, as the task is performed, records the number of variables. With this the robot is able to repeat the correct procedure automatically, ie *manual training*. This is the only way to proceed with the types of operational robots available at present.
2. The operator describes the task in terms comprehensible to the robot's computer which will then generate the necessary planning and the trajectories arising from this description. 'Artificial intelligence' may not be relevant here – if the operator uses FORTRAN (corresponding to a series of subprograms which then become established) then this is not a case of artificial intelligence; on the other hand, if a natural (oral) language is used, the complexity of processing allows legitimate reference to artificial intelligence (see Volume 2).

If a robot is equipped with well-adapted external, absolute sensors, and if the task is not completely specified, then *self training* can be used. The robot makes use of data and does not rely on training information. The robot records its successes and failures and changes its planning according to the nature of the task to be performed. In the absence of such sensors, a type of training can be considered which makes use of more advanced methods than manual instruction. In this, the task is not completely specified; the robot performs the task and then the 'teacher' indicates the necessary corrections. However, this type of training is of theoretical interest only.[58]

9.1 Methods of recording trajectories

An operator may steer a robot through the manoeuvres necessary to perform a task either *passively* (holding the end effectors, for example) or *manually* via a master station (see Figures 59 and 60). In the first case, information concerning successive configurations is recorded,

Figure 59. *Passive drive of a robot by an operator*

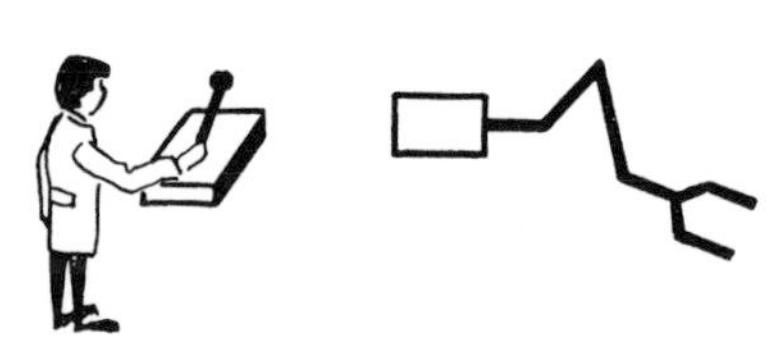

Figure 60. *Manual control from a master station*

ie values of articulated variables provided by the sensors. This information must then be converted into servo-control voltages (by analogue or numerical processing) before the robot can run in the automatic mode. In manual control the servo-control voltages are available during the training procedure and the robot is therefore *active*. (If a direct tape-recording is used, only manual control can be used.)

9.1.1 PERMANENT CONTINUOUS RECORDING

The servo-control input voltages to the motors during manual operation can be recorded on magnetic tape, eg for a robot processing six DOF and an end effector gripper which is closed, seven tracks are recorded (unless multiplex is used). When the task has been completed (and a recording made) the operation can be rerun in the automatic mode by replaying the recording – it should be noted that the gains at the recorder outputs must be correctly set.

9.1.2 PERMANENT DISCRETE RECORDING

The same type of recording can be obtained using a computer with a sufficient memory. The servo-control input voltages are sampled, digitally processed and then memorized. During operation the reverse operation occurs by way of an initialization process (see Figures 61 and 62).

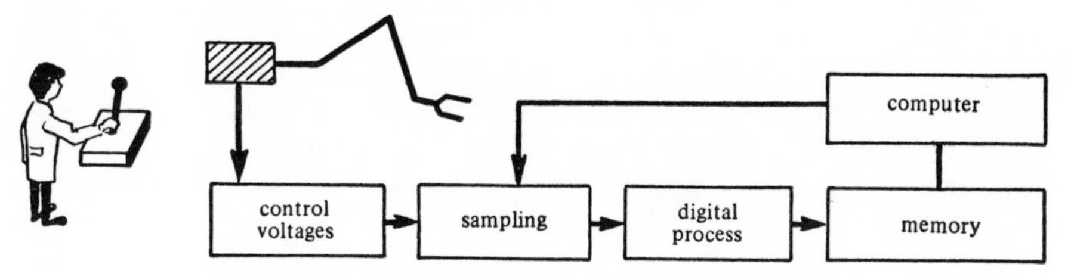

Figure 61. *Digital recording using an active robot*

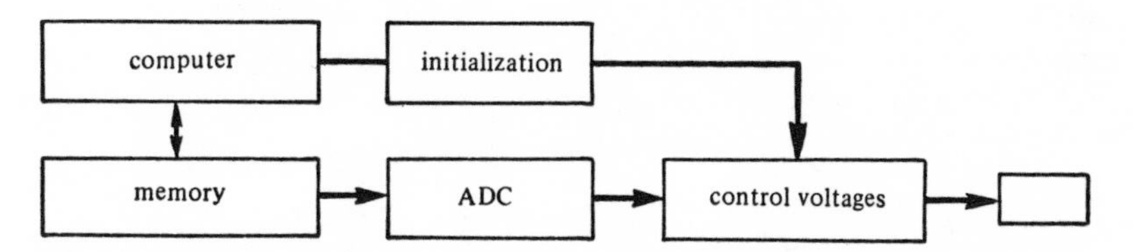

Figure 62. *Task execution in the automatic mode (training with an active or passive robot)*

When a robot is passive during the learning process, the data obtained from the sensors must be converted into control voltages (see Figure 63).

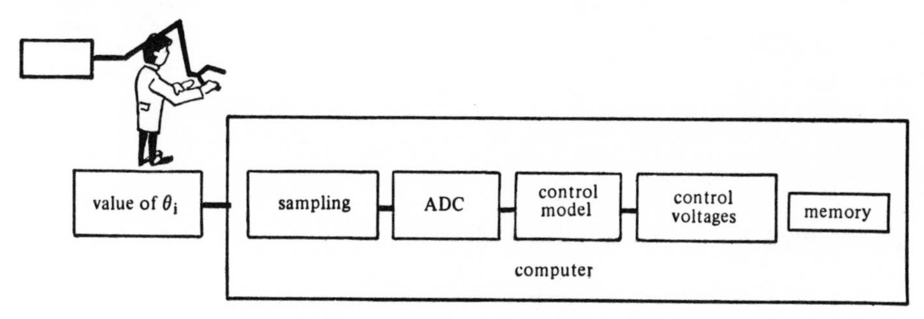

Figure 63. *Recording with a passive robot*

9.1.3 RECORDING SPECIAL CONFIGURATIONS

Consider the simple routine of picking up an object and placing it in a different location in the task space, whilst avoiding any obstacles. Elementary analysis of such a process indicates there are two types of configuration. The first are *precise* (not a synonym for 'single' – several configurations may be possible) configuration, eg the 'grasping' configuration used before the closing of a gripper or the 'deposit' configuration used before the opening of a gripper. Other configurations are *optional*, eg displacement configurations; the essential requirement is that the transition must be from one precise configuration to another. The computer will then generate trajectories between recorded configurations.

In this procedural mode, the computer must satisfy constraints which are related to the manipulated object eg maintaining the gripper in the horizontal. Alternatively, the computer could satisfy constraints which are related to the environment, eg the presence of obstacles.

In the first case the position of each obstacle must be described to the computer before an appropriate trajectory (which will allow movement without collision) can be generated. In the latter, obstacles are avoided without the need to inform the robot of their position (see Figure 64). This is achieved by dealing with the family of configurations arising from the required trajectory. In this way, trajectories between stored configurations can be based on one algorithm, eg interpolation.

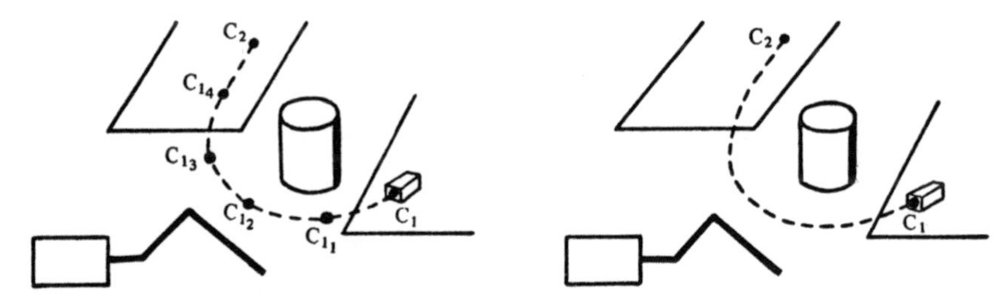

C_{1_i}: points corresponding to the configurations needed to avoid the obstacle. The obstacle is not described to the computer.

C_1, C_2: points corresponding to the training configurations if the obstacle is described to the computer and taken into account in the generation of trajectory.

Figure 64. *Recording training configurations and the performance of the associated trajectory generator*

9.2 Manual control used outside of training

9.2.1 MANUAL DRIVE OF THE ROBOT BODY

This method is used in applications such as paint-spraying and consists of the operator steering the robot through the routine. The trajectory (in practice the angular values of the articulations) is recorded and can be reproduced in the automatic mode as many times as required. The starting command is obtained from a relay which is triggered by the objects to be sprayed as they move on a production line.

9.2.2 USE OF A REMOTE MANIPULATION SYSTEM

A remote manipulation system (RMS) is an AMS which is handled by an operator. It links the movements of the operator with those of the robot (through a mechanical, electronic or computer coordinate 'transformer'). An RMS may have a variety of forms, depending on its application. There are three main types:

1. The 'classical' master-arm, as used in master-slave teleoperation: In this form, the structure of the RMS is identical to that of an ordinary robot except that the gripper is replaced by a handle (see Figure 65). This type of system is used in telemanipulation as it leads to symmetry between the master-arm and the slave-arm. This facilitates manipulation – when the master-arm is pushed forward, the slave-arm moves forward accordingly. Variable gain can be obtained between rotation of the master articulation and the corresponding slave articulation. In some models positional shift is possible; the 'whole' slave is moved relative to its original position. Such systems are used when work needs to be done in hostile environments, eg nuclear plants, underwater, foundries, etc. These applications will be considered in detail in Volume 3 of the series.

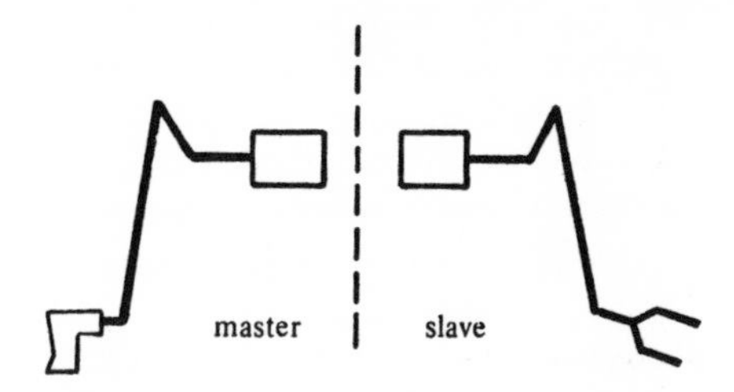

Figure 65. *The master-slave couple*

2. The 'puppet': This is a light structure, often possessing the same number of DOF as the robot, which can easily be manipulated. The learning process is started with the aid of such a device, and the robot thus records the necessary trajectories. This structure is used in such applications as paint-spraying.

3. The joystick: This device has the same number of DOF as the robot. The operator can guide the end effector of the robot using a joystick – if it is pushed forward, the gripper moves forward; if it is pulled back, the gripper moves backward; manipulations in other directions (left and right, up and down) can also be achieved using a joystick (see Figure 66).

If the handle of the joystick has a pressure control the movements can be produced without having to move the handle.

In all of these mechanical control systems, movements of the master must be transformed into movements of the robot. An important consideration is one of ergonomics – how efficiently the master translates the operator's instructions to the slave.

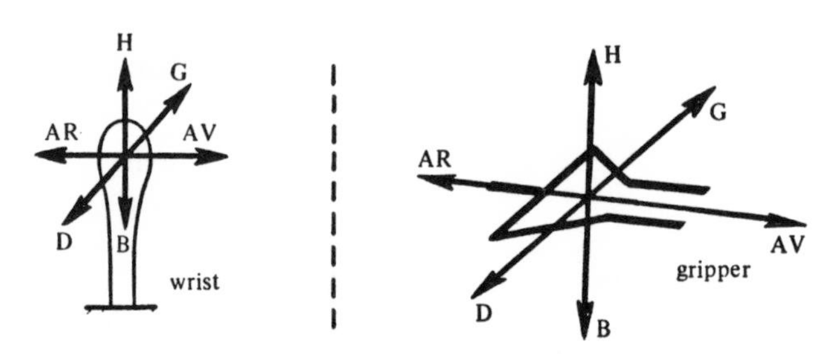

Figure 66. *Joystick control*

9.2.3 PUSH-BUTTON CONTROL

This is the oldest type of control, and is the most difficult to use. In it, each DOF is controlled by a push-button. The operator must decide on the best coordinate transformation: 'If I press button 3, angle θ_3 increases and the gripper moves forward'. There are also manipulators with mixed control – in some cases the arm is controlled by hand but the end effector is controlled by a push-button. In more sophisticated systems the buttons initiate orders 'move the gripper down, up, right, left, etc'. In some systems the longer the button is pressed, the faster the robot moves (speed control).

9.3 Improved controls

All the previously defined modes of control (except manual drive) are used not only for training purposes, but also in the effective control of the slave – particularly in teleoperations. A special application is in devices designed to help the handicapped.[59] Some handicapped people have some degree of control of some parts of their body and the signals obtained from these parts are coded and used as control signals. For example, myographic signals are derived from muscles in the nose, forehead, ears and from movements of the eyelids, mouth, etc.

For training purposes and in direct control, it is useful that a special function is required to perform a specific task. Some examples are: maintenance of the horizontal of a gripper, maintenance of an end effector parallel with its original orientation, movement of an end effector through an angle which is related to the vertical. On the other hand, the feedback which passes through a system can be useful as it allows the operator to determine the forces exerted on the system by the load, and whether there are any obstacles present. Finally, the facility that allows robots to operate at high speeds from a recording made at low speed has many important applications.

9.4 Trajectory generation

Suppose that a robot needs to attain two configurations (C_1 and C_2) to be able to perform a task and that these have been stored in the learning phase. The trajectory that connects C_1 and C_2 must be generated.

9.4.1 MODELLING AND TIMING PROBLEMS

Even a well-trained operator cannot control dynamic phenomena in the training phase. This may be because the system is not designed to do this (as in push-button control) or because only specific configurations are recorded when the robot is static. During the automatic execution of a task it is necessary to accomplish either exact time minimization or an overall minimization of the time required to complete the task. Because, for reasons discussed in Chapter 8, dynamic control is difficult to achieve in practice, kinematic or geometric models are not used for the control algorithm.

9.4.2 CONSTRAINTS

There are an unlimited number of ways of achieving the movement C_1 ($\underline{\Theta}_1$) to C_2 ($\underline{\Theta}_2$). There are, however, several constraints which need to be considered:

1. Constraints associated with the task space, such as the presence of obstacles and the maintenance of the horizontal of the gripper.
2. Constraints associated with the problems of control (often a consequence of constraints to the task space), such as the need to avoid singular configurations, the limits of the reach of the end effector, and the need to sample the trajectory in a way which is compatible with kinematic control.
3. Time constraints: in general, the shorter the trajectory the shorter the transit time through $\underline{\Theta}$ (since the trajectory can be generated off-line, before the task is started, the speed of execution of the task is not affected).

9.4.3 GENERATION OF ARTICULATED VARIABLES WITHOUT SPATIAL CONSTRAINTS

When there is no spatial constraint between C_1 and C_2 the simplest method of generating articulated variables (with respect to $\underline{\Theta}$) is as follows. First an increase, $\Delta\theta_{i\,\min}^{\max} = \delta$, is chosen which represents the smallest of the individual increases in the angle, θ_i, in the time interval, T (the control sampling period), which is compatible with a kinematic model. Consider the robot shown in Figure 50. Taking into account the leverages, masses and inertias during T, it is possible to produce a variation in θ_5 (for example, greater than the variation in θ_2) without invalidating the kinematic model. All the increments (θ_j) involved in the movement, C_1 to C_2, can be measured using values of δ. These variations must be compatible with the limits of articulation and their values must be determined. Figure 67 shows that to achieve the movement θ_i^1 to θ_i^2 it is necessary to pass through zero.

With this precaution in mind, consider a system with six rotational DOF. The following can be written:

$$\begin{aligned} \theta_1^1 - \theta_1^2 &= N_1 \delta \\ \vdots \quad & \quad \vdots \\ \theta_6^1 - \theta_6^2 &= N_6 \delta \end{aligned} \tag{9-1}$$

Suppose that the greatest value of N_i is N_4 (the number of increments in going from C_1 to C_2) for each variable, θ_i. The angular variation for each value of θ_i can be performed using increments of amplitude:

$$\begin{aligned} \Delta\theta_1 &= N_1 \delta / N_4 \\ \Delta\theta_2 &= N_2 \delta / N_4 \\ \vdots \quad & \quad \vdots \\ \Delta\theta_4 &= \delta \\ \Delta\theta_6 &= N_6 \delta / N_4 \end{aligned} \tag{9-2}$$

If the configuration C_1 ($\underline{\Theta}_1$) corresponds to the control vector $\underline{V}_1$ and $\underline{\Theta}_1 + \underline{\Delta\Theta}_1$ corresponds to $\underline{V}_1 + \underline{\Delta V}_1$, then:

Successive configurations	Values of each articulated variable for every configuration	Control values of each actuator for each configuration	Successive control vectors
$\underline{\Theta}_1$			$\underline{V}_1$
$\underline{\Theta}_1 + \underline{\Delta\Theta}_1$	$\underline{\Theta}_1 + N_1\delta/N_4$ $\underline{\Theta}_2 + N_2\delta/N_4$ $\vdots$ $\underline{\Theta}_4 + \delta$ $\vdots$ $\underline{\Theta}_6 + N_6\delta/N_4$	$V_1^1 + \Delta V_1^1$ $V_2^1 + \Delta V_2^1$ $\vdots$ $V_4^1 + \Delta V_4^1$ $\vdots$ $V_6^1 + \Delta V_6^1$	$\underline{V}_1 + \underline{\Delta V}_1$
$\underline{\Theta}_1 + 2\,\underline{\Delta\Theta}_1$	$\underline{\Theta}_1 + 2N_1\delta/N_4$ $\vdots$	$V_1^1 + 2\Delta V_1^1$ $\vdots$	$\underline{V}_1 + 2\,\underline{\Delta V}_1$
$\vdots$	$\vdots$	$\ddots$	$\vdots$
$\underline{\Theta}_1 + N_4\underline{\Delta\Theta}_1 = \underline{\Theta}_2$	$\underline{\Theta}_1 + N_1\delta$ $\vdots$	$V_1^1 + N_4\Delta V_1^1$ $\vdots$	$\underline{V}_1 + N_4\underline{\Delta V}_1 = \underline{V}_2$

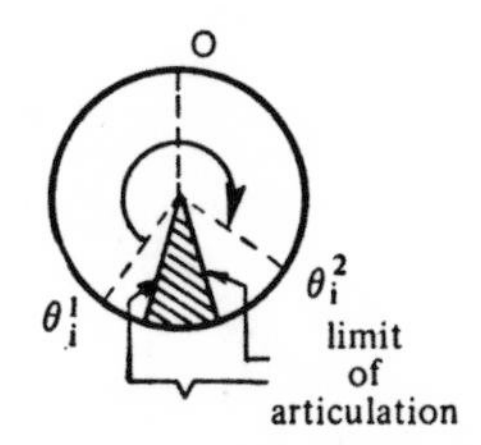

Figure 67.

The sequence of the control vectors, $\underline{V} = \underline{V}_1 + k\underline{\Delta V}_1$, can then be recorded. For systems in which this kind of interpolation can be used it is possible to impose time, t_1^2, between two configurations. In addition to the number of increments in θ_i, the sampling period (T) can be computed. However, if the time is too short (and thus incompatible with the control capacity or kinematic hypotheses) t_1^2 must be increased to allow computation.

9.4.4 GENERATION OF ARTICULATED VARIABLES FOR A SPECIFIC FINAL CONFIGURATION

Problems can arise in the generation of articulated variables which are associated with the final configuration. 'Intermediate' configurations (eg the configurations C_{1_1} and C_{1_2} in Figure 64) are usually achieved with poor precision and at speed. However, final configurations (C_2 in Figure 64) must be precise and achieved without overshoot or oscillation. For the situation described in section 9.4.3 the speed of the articulated variable is theoretically constant ($\theta_i = N_i\delta/N_4T$). It is evident that the change in $\underline{\Delta\Theta}$ producing configuration C_2 (using open-loop control) will entail inertial effects which may produce overshoot. Using closed-loop control, however, the final configuration can be attained with precision, but possibly with oscillation. This problem can be overcome by reducing the speed of the robot as it approaches its final configuration, ie the amplitude of $\Delta\theta_i$ (in the sampling period, T) decreases as the final configuration is approached.

There are many ways in which the speed can be reduced – all involve a decrease in the value of $\Delta\theta_i$ produced by a braking system. Hill[60] and Khalil[61] have proposed operational methods which involve a control of angular variations. This allows a time variation in θ_i of quasi-sinusoidal form. Acceleration is followed by deceleration to achieve a braking effect (see Figure 68).

If the sampling period (T) is fixed, a suitable choice of N should be made heuristically, eg by determining the movement of the end effector during time T, or alternatively, from the speed of movement of the end effector, which is compatible with the kinematic model. For example, if T = 20 ms, a movement of the end effector of 1 cm produces a speed of 50 cm/s. The choice of N is, therefore, determined by the structure of the robot.

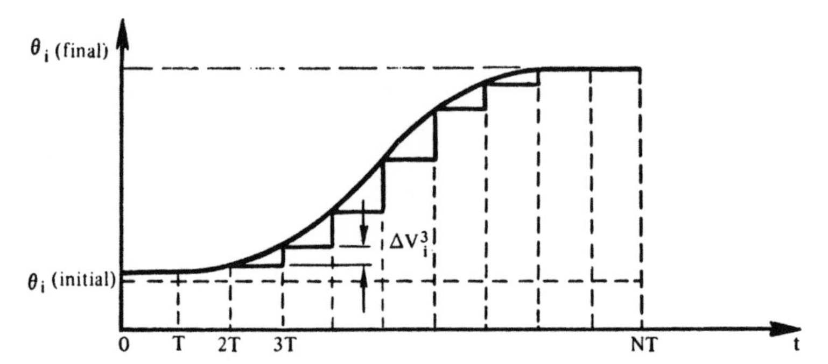

Figure 68. *The change of the value of an articulated variable with time, showing acceleration followed by braking*

Another method consists of dividing the interval, $\theta_{i\,(\text{initial})}$ to $\theta_{i\,(\text{final})}$, into N equal intervals.[62] Then from stage L (<N) (see Figure 69) every interval can be further divided into n intervals. At the last stage of N, the interval is thus divided into nx stages. The efficiency of deceleration can be improved further by dividing stage N in such a way that every new segment has a smaller amplitude than the previous one. For example:

$$\Delta\theta_i^N/n^2,\ \Delta\theta_i^N/n^3,\ \Delta\theta_i^N/n^4,\ \text{etc.}$$

until $\theta_{i\,(\text{final})}$ is at a distance less than ϵ. Stage N can then be replaced by ζ stages:

$$\zeta = [1 + (\log\ \Delta\theta_i^N/\epsilon\ )/\log n]$$

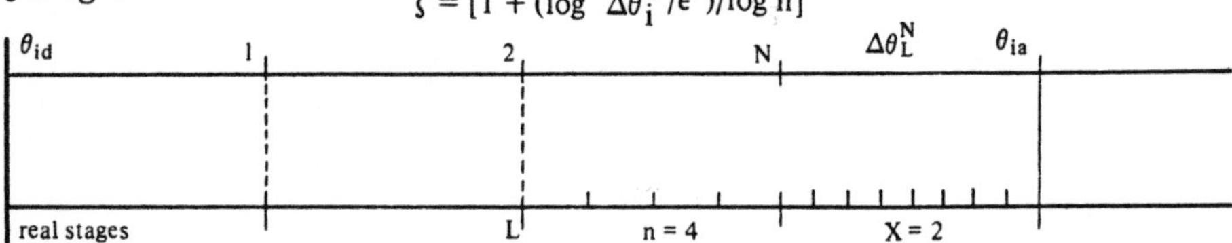

Figure 69. *A braking method*

9.4.5 GENERATION OF TRAJECTORIES WITH CONSTRAINTS IN THE TASK SPACE

Apart from time constraints, the spatial specifications of the end effector may be such that it must follow a straight line or a circle or some other specified curve; or that the gripper must remain in the horizontal or turn about an axis in a pre-determined direction.

These constraints amount to a precise specification of the trajectories of some points in the task space (R_0), or the orientation of the coordinates (as a function of the position in R_0) of some specified point.

The problems associated with these constraints have been discussed in Chapter 5.

In practice, spatial constraints are generally applied to the end effector. Take the example of a gripper having two fingers; the constraints can be defined with a knowledge of:

1. The position of its centre of gravity (G) in the reference coordinate set (R_0): X_G^0, Y_G^0 and Z_G^0.
2. The orientation of the coordinate set linked to the gripper at G with reference to R_0. The unity vectors of the three axes of this coordinate set can be written as ($\underline{a}, \underline{n}, \underline{s}$) (see Figure 70).

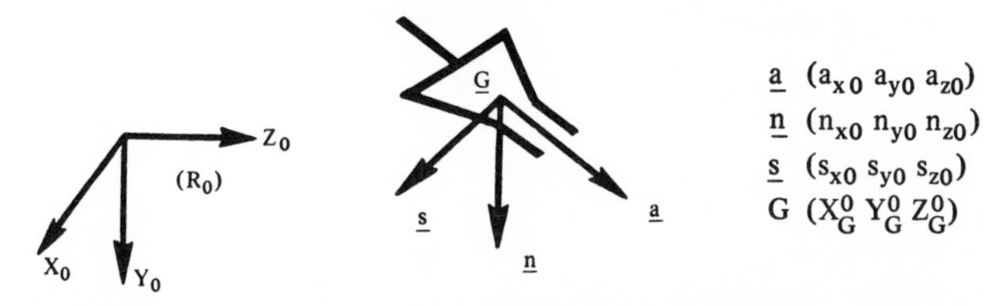

Figure 70. *The positioning and orientation of a two-fingered gripper*

In practice, every spatial constraint can be expressed by relationships imposed on some of the 12 components of $\underline{a}, \underline{n}, \underline{s}$ and G. For example:

1. To maintain the horizontal of the gripper: $n_{y0} = +1$ (or $a_{y0} = s_{y0} = 0$).
2. To maintain the vertical of the gripper: $a_{y0} = 1$ (or $n_{y0} = s_{y0} = 0$).
3. To maintain the same direction for the vector $\underline{a}$ in R_0: $a_{x0} = \alpha_1$ and $a_{y0} = \alpha_2$ (since $|a| = 1$).
4. To direct the gripper permanently (ie the vector $\underline{a}$) towards a fixed point (P) in R_0 [if G and P in R_0 are known the cosines directrix of the line GP in R_0 can be calculated – these are $\beta_1(t)$, $\beta_2(t)$ and $\beta_3(t)$]: $a_{x0}(t) = \beta_1(t)$, $a_{y0}(t) = \beta_2(t)$ and $a_{z0}(t) = \beta_3(t)$.
5. To make the gripper follow a circumference whilst maintaining it as normal to the circular plane (see Figure 71).

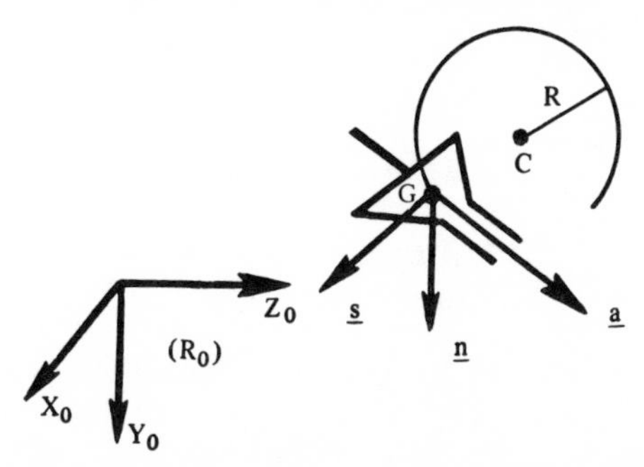

Figure 71. *Circumference tracking of a gripper*

Suppose that the circle is defined in the set of coordinate axes R_0 by the position of its centre C (X_C, Y_C and Z_C), its radius (R), and the cosines directrix of its plane (cos 1, cos 2). For a circle with a radius, R, centred on C, the following equation can be written:

$$(X-X_C)^2 + (Y-Y_C)^2 + (Z-Z_C)^2 = R^2 \quad (9\text{-}3)$$

The equation of the plane has the form:

$$(X-X_C)\cos 1 + (Y-Y_C)\cos 2 + (Z-Z_C)\cos 3 = 3 \quad (9\text{-}3b)$$

and the intersection of the plane and circle can be written as:

$$(X_G-X_C)^2\left[1+\frac{\cos^2 1}{\cos^2 3}\right] + (Y_G-Y_C)^2\left[1+\frac{\cos^2 2}{\cos^2 3}\right] + \left[2\frac{\cos 1\cos 2}{\cos^2 3}\right](X_G-X_C)(Y_G-Y_C) = R^2 \quad (9\text{-}4)$$

$$Z_G - Z_C = [-\cos 1\,(X_G-X_C) + \cos 2\,(Y_G-Y_C)]/\cos 3 \quad (9\text{-}5)$$

Movement of the position of G must satisfy equations (9-4) and (9-5). Moreover, the fact that $\underline{a}$ is perpendicular to the circle leads to the supplementary constraint:

$$a_{x0}\cos 1 + a_{y0}\cos 2 + a_{z0}\cos 3 = 1 \quad (9\text{-}6)$$

$$\text{with: } \cos^2 1 + \cos^2 2 + \cos^2 3 = 1 \quad (9\text{-}7)$$

$$a^2_{x0} + a^2_{y0} + a^2_{z0} = 1 \quad (9\text{-}8)$$

9.4.6 GENERATION OF INCOMPLETELY SPECIFIED TRAJECTORIES

From the points discussed in section 9.4.5 two cases emerge:

1. The number of constraints is sufficient to determine a unique trajectory. Sampling should establish relationships between the various increments $\Delta\theta_i$.
2. The number of constraints allows some choice of trajectory (eg movement from C_1 to C_2 while maintaining the horizontal of the gripper). Under such conditions how is the trajectory chosen and how is it sampled?

Consider the previous example in which C_1 ($\underline{\Theta}_1$) and C_2 ($\underline{\Theta}_2$) are known and the constraints impose $a_{y0} = s_{y0} = 0$. Assuming that the robot is similar to the one illustrated in Figure 50:

$$s_{y0} = 0 = S1C4C6 + C1C(2+3)S4C6 - S1S4C5S6 + C1C(2+3)\,C4C5S6 - C1C(2+3)\,S5S6 \quad (9\text{-}9)$$

$$a_{y0} = 0 = S1S4S5 - C1C(2+3)C4S5 - C1S(2+3)\,C5 \quad (9\text{-}10)$$

from which the following equations can be derived:

$$\sum_{i=1}^{6} F_i(\underline{\Theta})\,.\,\Delta\theta_i = 0 \quad (9\text{-}11)$$

$$\sum_{i=1}^{6} g_i(\Theta) . \Delta\theta_i = 0 \qquad (9\text{-}12)$$

$$\begin{aligned} \theta_6^2 - \theta_6^1 &= \delta\theta_6 = N_6\delta \\ \theta_5^2 - \theta_5^1 &= \delta\theta_5 = N_5\delta \\ &\vdots \\ \theta_1^2 - \theta_1^1 &= \delta\theta_1 = N_1\delta \end{aligned} \qquad (9\text{-}13)$$

The most significant variation ($\delta\theta_i$) in system (9-13) can be divided into N_j increases of $\delta\theta_i/N_j$, compatible with the other characteristics of control. Equations (9-11) and (9-12) provide two relationships which involve six increases; therefore, increases can be imposed in four variables. Suppose that θ_4 has the greatest variation, then:

$$\Delta\theta_1 = N_1\delta/N_4 \qquad \Delta\theta_2 = N\theta_2 = N_2\delta/N_4$$
$$\Delta\theta_3 = N_3\delta/N_4 \qquad \Delta\theta_4 = \delta$$

Using equations (9-11) and (9-12) it is possible to determine $\Delta\theta_5$ and $\Delta\theta_6$ (which will be constant). However, this procedure is concerned with the possible transition through singularities. Computation indicates that $\Delta\theta_5$ and $\Delta\theta_6$ have unlimited values and suggests that the end stops for θ_5 and θ_6 meet. In both the cases described here the method consists of choosing four of the six variables (eg $\Delta\theta_2$, $\Delta\theta_4$, $\Delta\theta_5$ and $\Delta\theta_6$) and calculating the other two using equations (9-11) and (9-12). The following can then be determined:

1. Computation of the greatest excursion, $\theta_i^2 - \theta_i^1$.
2. A suitable choice of N_j.
3. A suitable choice of the four angles on which the following can be imposed:

$$\Delta\theta_K = (\theta_K^2 - \theta_K^1)/N_j$$

4. Computation of the variations in the angles $\Delta\theta_p$ and $\Delta\theta_q$ [using equations (9-11) and (9-12)].

As an other example, suppose that the constraints impose the position of G but not the orientation of the end effector. The trajectory can then be defined in the task space using an equation which relates X_G, Y_G and Z_G. This trajectory can then be sampled for X, eg by defining ΔX_G, and the corresponding values of ΔY_G and ΔZ_G are indicated by the constraints. To pass to $\Delta\theta$ the methods developed in Chapter 6 may be used. The use of:

$$\underline{\Delta\Theta}^* = J^+ \underline{\Delta X} + (J^+J - \Uparrow)\underline{z} \qquad (6\text{-}64)$$

allows $\Sigma\Delta\theta_i^2$ to be minimized and using a particular choice of the function from which $\underline{z}$ is derived, the distance from the articulated limits can be minimized.

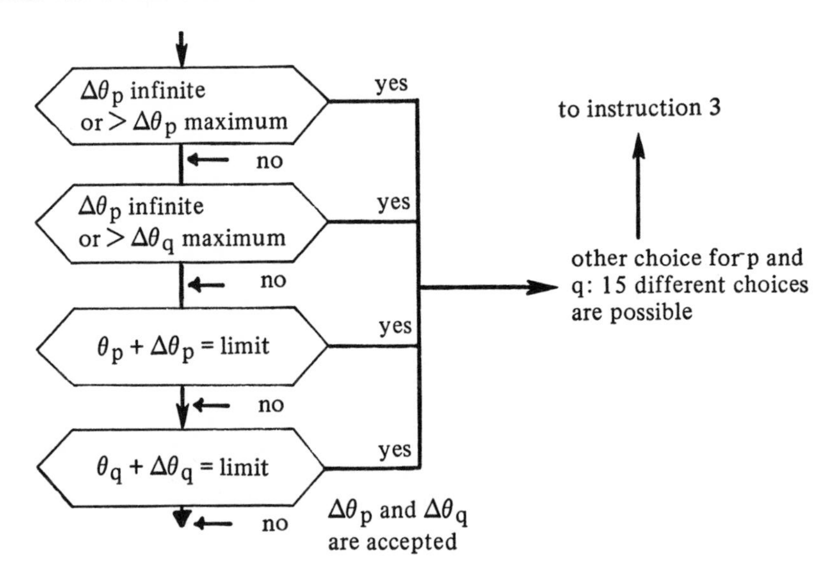

Conclusions

Although computer training of robots having different computational devices may take different forms the sample principle is always used – the recording of articulated variables for configurations central to the execution of a particular task. The trajectory linking these configurations can be generated using several interpolations. In the training mode it is not the operator who determines the recorded configurations. Rather, the operator determines the configurations which arise from the execution of the task (and their associated trajectories) using automatic examination of the environment and *a priori* data. This infers exploitation of artificial intelligence techniques rather than trajectory generation.

Chapter 10

Tasks and performance of articulated robots

Although a task can be described accurately, in practice it is difficult to achieve precision. There are only a limited number of factors that can be accurately specified. These include the average and maximum operational speeds, positional precision, useful load and attainable reach. It is clear that versatility, cost, sensitivity to external factors and other characteristics should be taken into account when assessing performance. With this in mind, it would be useful to consider the elements which should be considered when purchasing a robotic system.

10.1 Description of tasks

10.1.1 DEFINITION OF A TASK

In Chapter 1 it was shown that a robot is intended to act on its environment by performing a transformation of its 'universe'. A task can therefore be considered as the difference which exists between two stable states of such a universe. Assuming that the task is compatible with the capacity of the robot, modification of its universe can be achieved in an apparently unlimited number of ways. However, modification can never be spontaneous. The meaning of the word 'task' becomes obscure as not only does it imply the difference between the initial and final states of a universe, but also the means of carrying out the work necessary to bring this about.

10.1.2 WHY DESCRIBE A TASK?

Task description is justified for three reasons:

1. The need to be able to communicate with the robot: Orders can be oral or transmitted via a computer. The robot must be able to understand such orders and have access to all the data necessary for their execution. In simple robots (those that do not interreact with their environment) these data are stored in the memory of the robot's computer and are, therefore, readily available. These data constitute the *task description* and can be used to generate all the movements necessary for the execution of a particular task. A robot performs a task (grasping, carrying and placing an object, for instance) using an infinite number of sequential actions. A task description thus contains variants and it is essential that there are sufficient elements (*pertinent parameters*) to allow *induction* of the correct sequence of movements for the execution of the task. Therefore, there are a number of modes of execution for every task; the choice of execution mode is influenced by various optimization criteria. This 'necessary and sufficient element' condition is important. In every task description it is important that redundancy of information is suppressed (ie sufficient elements) but that nothing is suppressed which might be of consequence to the robot (ie necessary elements); in practice this is very difficult.

Consider the situation of pattern recognition from optical images. The information to be extracted is distributed over a number of points which, together, describe the task. For example, an object might be blue and have a length of 10 cm, its shape might describe a right angle, it might be positioned in a particular location with a particular orientation, it might be necessary to turn the object through a particular angle, etc. In this multiplicity of points it is essential that those containing maximum information are combined and interpreted before the movements of the robot are determined.

2. The need to be able to simulate execution of a task: A computer-controlled robot is a very complex system. Modern computer techniques allow the behaviour of a robot to be simulated and, therefore, an appraisal of its suitability for a particular task can be made. However, simulation is possible only if a task can be described in computer language.

3. The need to be able to design a robot suitably adapted to a particular task: Using various specifications and a suitable task description, the computer-aided design of a robot becomes possible.

10.1.3 ASPECTS OF THE DESCRIPTION

Task description involves two distinct aspects which, in the past, have been a source of some confusion:

1. The *fundamental* aspect involving 'What are the elements of a task description and how must they be organized to initiate task execution?'.
2. The *formal* aspects involving 'Which is the most suitable language for their description?'.

It has often been the case that the formal aspects have been considered without allowance for the fundamental aspects. This has resulted in a proliferation of computer languages, with varying degrees of usefulness.

10.1.4 CHARACTERISTICS OF LANGUAGES IN USE AT PRESENT

There are two main types of language:

1. Control or programming languages.
2. Languages derived from machine-tool control or from automatic machines of a production line.

The first type is usually specific to one type of computer-controlled robot, eg WAVE and REX are used at the Jet Propulsion Laboratory, DIAL at Stanford University, STRIP and NOAH at Stanford Research Institute, BUILD and LAMA at the Massachusetts Institute of Technology, AUTOPASS and EMILY at IBM, VAL at Unimation, SIGLA at Olivetti, RAPT at Edinburgh University and LAMA-S at the Spartacus Laboratory.[63]

Each of these languages is different and, although generalized, there are problems associated with their use. Generalized languages must offer 'primitives', as they are called, which can be used to compile a hierarchical set of programs for use in the generation of sequences of control signals. The use of such a language implies a knowledge of the constraints of a system and the existence of a task description. It is difficult to transfer even simple orders (such as 'lift' and 'fetch') from one system to another without a change of language – even though the same type of computer hardware may be in use in both systems. A language which is used in one type of information-processing device can be used in another, although the primitives will not have the same meaning in both systems. Re-implementation is difficult because all the new low-level (control) programs will be invalid in the new system.

Languages relevant to geometrical modelling and adapted to machine tools can more easily be used because all machine tools are similar and the tasks they perform fall within a comparatively narrow field. Therefore, low-level primitives can be applied universally and allow the determination of such variables as length, mass, physical properties of materials and cutting speeds, etc.

Languages which are used for such applications include PRONO, APT, PADL1 and PADL2.[64] The problems associated with the control languages of machine tools and robots will probably develop in parallel with the development of problem analysis and task analysis.

10.1.5 A POSSIBLE TASK DESCRIPTION FOR AN ARTICULATED ROBOT

A four-level description will be described here which is similar to that proposed by Latcombe.[65]

10.1.5.1 The first level: the goal task

Definition: The first descriptive level is called the objective task or goal task and fulfills the task definition described in section 10.1.1, ie the difference which exists between two stable states of the robot's universe. It should be noted that this is not directly concerned with the robot, rather the goal task concerns a specific description of the robot's universe because:

1. For a robot in a fixed position the universe under consideration does not extend beyond the attainable reach of the robot's end effector(s).
2. The description of the universe is a function of the task to be performed.
3. The description is not exhaustive since some of the elements present in the 'reachable' space (either temporary or permanent) are not involved in the execution of the goal task. However, other elements (apparently permanent and independent of the initial and final states) need to be taken into account as they act as constraints or perturbations of the goal task.
4. The goal task description must be inductive – this allows a description of other task levels.

Elements involved in the description of a goal task: That the goal task must be inductive emphasizes the need for a 'closed loop'. To determine the elements in the description of a goal task it is necessary to analyse what a particular robot can do. Examination of the properties of a robot possessing an end effector or working tool allows a generalization that a robot is capable of applying forces and carrying out movements within a specified space. In an industrial context robots are capable of maintaining objects, assembling parts, and carrying out fitting operations and inspection procedures. In addition, they can change the nature of an object by contact (soldering, cutting, etc) or from a distance (paint spraying). The characteristics of articulated robots allow the robot's universe to be divided into:

1. Objects of interest, ie those on which the robot is to act.
2. Other objects.
3. The environmental medium, ie the properties of the medium in which the work is being carried out (underwater, at high temperature, under vacuum, etc).

Objects of interest: These can be divided into four categories:

1. The initial state (IS): This can be described in terms of two groups of elements, namely those which concern (i) the description of objects, independent of their spatial situation (ie appearance in terms of shape, colour, group specification, etc) – IS_1; and (ii) the initial spatial situation (ie orientation of the object relative to a reference coordinate set of axes, its centre of gravity, etc) – IS_2.

2. *The evolutive or transitory state (ES):* This concerns the spatial evolution of the object and the ways in which it changes as a required transformation occurs (ie task execution). It is particularly relevant to IS_2 (t), for example, when the object of interest is on a conveyor belt. It is essential, therefore, that there is an adequate knowledge of IS_2 (t) to allow transformation in the 'open loop' (this emphasizes the need for externally sensed actuators).

3. *The permanent state (PS):* This encompasses the physical and mechanical properties of the object of interest – and must be determined before a required transformation is possible. The details include such characteristics as density, fragility, elasticity, etc. This is a permanent state since these characteristics are unlikely to change during task execution.

4. *The final state (FS):* This includes a description of (i) the objects of interest, using the parameters of IS_1; and (ii) their situation, using the parameters of IS_2 but with values determined after task execution.

For ρ objects of interest:

$$OI = (OI_1, OI_2, \ldots OI_\rho)$$

For an object, i:

$$OI_i = (IS_1^i, IS_2^i, ES^i, PS^i, FS_1^i, FS_2^i)$$

where $IS_1^i = (\text{name, shape, colour, dimensions} \ldots)_i$

$$IS_2^i = (X_{G_i}^0, Y_{G_i}^0, Z_{G_i}^0, \alpha_i^0, \beta_i^0, \gamma_i^0, \ldots)$$

$X_{G_i}^0$, $Y_{G_i}^0$ and $Z_{G_i}^0$ are the centres of gravity in the reference set

α_i^0, β_i^0 and γ_i^0 are the orientations of the object relative to the identified set

$$ES^i = ES_2^1(t) = [X_{G_i}^0(t), Y_{G_i}^0(t), \ldots \alpha_i^0(t)]$$

$$PS^i = (\text{density, fragility, elasticity})_i$$

$$FS_1^i = IS_1^i(t = \infty)$$

$$FS_2^i = IS_2^i(t = \infty)$$

Other objects: There are three types:

1. Those that might be included as an object of interest.
2. Objects which merge with obstacles that are geometrical, mobile objects – these can be described to a robot as prohibited zones.
3. Those produced by the environment, eg heat, light, radiation, smoke, etc.

It can be said that:

other objects = [prohibited zone l(t), . . . prohibited zone k(t)]

The properties of the environmental medium: If task description is used in the design of a robot the properties of the environmental medium become very important. However, without absolute external sensing the effect on *a priori* planning is minimal.

10.1.5.2 The second level: the tool task

This second level concerns the processes that make transformations possible. Tools of specific types are used depending on the specifications of the required transformation and the procedures that are available. It can be said that:

tool task = (process, tool 1, tool 2, . . . work norms, ie rotational speed, pressure, speed of movement, etc)

It should be noted that the tool task does not describe what the tools do, but rather the type of tool and the code of practice of its use. It concerns only a part of the data used in the control of machine tools.

10.1.5.3 The third level: the interaction task

This arises from the interaction between the robot and its environment and a knowledge of this is essential for the execution of tasks which would be impossible without such interaction. This describes parameters which were not considered at the second level.

As a practical example, consider the assembly of two metallic components – a cylinder which is to be introduced into a straight bore (the tolerance involved here is quite severe). Does a knowledge of the goal task provide all the necessary geometrical and physical information?

The tool task description states that if the tool used is a gripper then the operation rules are reduced to the available gripping force. It is evident that control of the speed and position is not always sufficient during the precise assembly of components. The forces acting between the cylinder and the bore must be controlled. The interaction task indicates that, in the present example, the reaction forces between the cylinder and bore should be set to produce a given direction and value. Task interaction does not arise merely from the fact that a robot is controlled in the 'open loop'.

10.1.5.4 The fourth level: the robot task

The descriptions of the previous levels allow the necessary trajectories to be generated. Generation in the task space, the articulated variable space and the control variable space constitute the robot task. As discussed previously (when training techniques were considered), useful trajectories are characterized by a number of points on the trajectory at which the configuration of the robot is crucial. The trajectory is sampled between two such successive points. There are numerous ways of describing crucial configurations,[66-68] including the Cartesian coordinates of a point on the end effector and the cosines directrix of its normal axes; the components of a vector and the angular position of the vector of the end effector in a known coordinate set; the Eulers' variables and others (see Chapter 6).

Fournier[69] described a method which allows the automatic step-wise generation of a robot trajectory between two crucial configurations. At any one moment a solid can be characterized by the speed of one of its points and its instantaneous rotational speed relative to a fixed reference point. If the position of the point and the orientation of the solid (in the initial state) are known, then the two vectors that constitute the kinematic rotator associated with the solid allow the trajectory to be determined:

$$\tau(0, S|R_0; R_0) = [\underline{\Omega}(S|R_0; R_0)|\underline{V}(0, S|R_0; R_0)] \qquad \textbf{(10-1)}$$

where τ is the kinematic rotator associated with the solid (S) expressed in a fixed set of reference coordinates (R_0); and Ω is the rotational speed vector of S relative to, and expressed in, R_0.

In the transition between two configurations a kinematic relationship is satisfied (i) totally, if the six components of the rotator (τ) are imposed; and (ii) partially, if some of the components of the rotator (τ) are fixed. Fournier was able to express all the movements of an end effector using the following six mechanical relationships:

Point contact (frictionless contact between a sphere and a plane):

$$\tau = [(\alpha, \beta, \gamma)^T |(u. \quad w)^T] \qquad \textbf{(10-2)}$$

Linear contact (frictionless contact between a cylinder and a plane):

$$\tau = [(0, \beta, \gamma)^T | (u, 0, w)^T] \qquad \textbf{(10-3)}$$

Surface contact (frictionless contact between two planes):

$$\tau = [(0, \beta, 0)^T | (u, 0, w)^T] \qquad \textbf{(10-4)}$$

Gutter (frictionless movement of a sphere inside a cylinder of identical diameter):

$$\tau = [(\alpha, \beta, \gamma)^T | (0, v, 0)^T] \qquad \textbf{(10-5)}$$

Swivel joint (frictionless contact between two concentric spheres of identical diameter):

$$\tau = [(\alpha, \beta, \gamma)^T | (0, 0, 0)^T] \qquad \textbf{(10-6)}$$

Sliding bolt (contact between two concentric cylinders of equal diameter):

$$\tau = [(0, 0, \gamma)^T | (0, 0, w)^T] \quad (10\text{-}7)$$

By combining these relationships:

The rotational relationship is:

$$\tau = [(0, \beta, 0)^T | (0, 0, 0)^T] \quad (10\text{-}8)$$

The prismatic relationship is:

$$\tau = [(0, 0, 0,)^T | (0, 0, w)^T] \quad (10\text{-}9)$$

The helical relationship is:

$$\tau = [(0, 0, \gamma)^T | (0, 0, \lambda \gamma)^T] \quad (10\text{-}10)$$

If the environmental contact reactions (defined in the interaction task description as a state reaction) are not considered then the kinematic rotator constitutes a complete description of the robot task.

10.1.6 CONCLUSIONS

Every task that is carried out by a robot can be described at four levels: goal task, tool task, interaction task and robot task. These levels are defined by the significant configurations and by the kinematic rotator that relates each of the configurations. This description is ideally suited for exploitation using computers (and also useful for continued adaptation, by extension, as a contraction of two-level descriptions).

10.2 The performance of articulated robots

Robotic equipment is designed with the specifications of its performance in mind. The performance of a robotic system can be defined by the constraints on its application. Such constraints include mass and power independence, accuracy, speed and reliability. An interest in the performance of teleoperation systems led Johnson[70] to devise an index of 32 classes of performance which gave several hundred indexes for a complete system. The number of parameters complicates the assessment of the capacity of a system.

There are several points concerning the performance of robots in practice which are of particular interest:

1. What can the robot do? – is it capable of performing the tasks required?
2. What technical constraints will the use of the robot impose? – space required, protection against dust, floor strengthening,

interface with the production line, need for safety devices, etc.
3. What does the use of the robot require of the operator? – programming, adaptability, etc.
4. What are the financial advantages of using the system? – outlay, profitability, etc.

10.2.1 TASK PERFORMANCE

The main characteristics of a robot which performs a task using an end effector can be described as:

1. Reachable space: Although this can simply be expressed as units of volume, the actual shape of the area researched by the end effector is of prime importance. This can be complex since it might be determined by the combination of several articulations. It can more simply be described in terms of the surface embedded in the real volume (the surface having a simple shape – the intersection of two spheres, ellipsoids or parallelepipeds, for example). Otherwise a graphical representation is required. Thus, there are two parameters: volume and shape.

2. Orientation of the end effector: The presence of effectors on each articulation and the possibility of mechanical coupling between the articulations means that it is not always possible to obtain the maximum range of all DOF of the end effector in the reachable space. The minimal angular ranges of the end effector relative to a fixed set of coordinate axes in the reachable space must be known. Alternatively, a map of these ranges must be established. This is more useful than a design specification which describes the range of each articulation relative to a fixed set of coordinates and neglects the problems associated with stops and couplings, thus limiting the potential application of the mechanism.

3. The payload: The torques developed by the actuators are a function of the configuration. Thus, the maximum payload cannot be dealt with in the whole of the reachable space nor in all possible orientations of the end effector. The maximum payload in the optimum configuration and the best overall payload (ie the load that can be carried in the whole of the reachable space) provide the best description of the payload – alternatively, a map of the payload as a function of configuration could be used.

4. Positional precision: There are three aspects which should be considered:

(a) With what precision can the final configuration be achieved? This is dependent on the precision of the tip of the arm and the centre of gravity (mm) and angular precision (radians) of the end effector. In the constant control mode, precision is dependent on the required zone of the reachable space and the payload – since elasticities may be significant near maximum payload. For this type of 'static' precision it is possible to state a load limit and construct a multidimensional map.

(b) Starting from a fixed position of the carrier and a fixed orientation of the end effector, what is the smallest realizable variation of these values? This factor of *sensitivity* is dependent on the type of control available and on the type of robot (and is of prime importance in automatic assembly applications). Sensitivity is a function of payload and configuration.

(c) What is the precision of repeated movement?

It would appear that in practice positional control is an important factor in robot performance, but that dynamic (speed) precision is not.

5. Speed: This is a fundamental characteristic for robots intended to be incorporated into a manufacturing system. The most important aspect here is the task-execution time. However, in an *a priori* evaluation this is not known and the situation is limited to a consideration of the translational and rotational speeds of the system.

Suppose that during a sampling period (T) there is a variation in $\underline{\Theta}_i$ of $\Delta\underline{\Theta}_i$. The positional and orientational variation of the end effector during this time is:

$$\Delta\underline{X} = J\,(\underline{\Theta})\,\Delta\underline{\Theta}$$

and the instantaneous speed in the task space is:

$$\dot{\underline{X}} \simeq (\Delta\underline{X}/T) = J(\underline{\Theta})\,(\Delta\underline{\Theta}/T) = J\,(\underline{\Theta})\cdot\dot{\underline{\Theta}} \qquad \textbf{(10-11)}$$

However, this is not so if $\dot{\underline{\Theta}}$ is constant as $\dot{\underline{X}}$ is dependent on the configuration ($\underline{\Theta}$). Manufacturers seem to evade the subject of speed performance; instead they describe the orders of magnitude or maximum speed of the translations of the end effector. Alternatively, they describe the rotational speed of the articulations, whereas it is the value of $\underline{X}$ (or preferably $|\dot{\underline{X}}|$) which is most useful to the user.

When assessing real speeds it is essential that:

(a) A distinction is made between arm speed (m/s) and the rotational speed of the end effector (rad/s), although these speeds may be coupled slightly.

(b) Data are available of execution times which can be used as standards.

(c) Typical values of vector $\underline{\Theta}$ and the matrix $J(\underline{\Theta})$ are available so that the appropriate value of $\underline{X}$ can be established for the transition through configurations of interest.

6. Reliability: This can be classed as a task performance and is expressed in terms of failure rate or percentage down-time[71] (the time during which a robot is not carrying out its assigned function). Reliability can be presented as a failure-frequency diagram. There are two main classes of failure:

(a) Complete functional breakdown.
(b) Degeneration of performance, eg spatial precision or loss of a DOF – this can be avoided by the use of a redundant DOF which, in an automatic process, allows for failure compensation. This aspect of reliability is now under investigation.[72]

7. *Synchronization with other machines:* In an industrial situation successive processes may be related in a time sequence. Robots must, therefore, be capable of working in synchrony with other machines and computers. It should be noted that the number of possible inputs and outputs is of importance when considering robot performance.

10.2.2 MANPOWER REQUIREMENTS

The main concern here is the use of a robot by an unskilled operator. The skills needed to make a robot work are associated with a knowledge of hardware (servo-control adjustment) and software (programming). It is useful to know if, after a short period of training, an unskilled operator would be able to control a robot system. Also, whether repairs could be carried out by the user or whether they need to be done by the manufacturer?

10.2.3 PERFORMANCE ECONOMICS

The cost effectiveness of an industrial robot is dependent on:

1. The cost of investment: purchase price, cost of initiation or modification of an existing system, etc.
2. The running expenses: energy costs, maintenance, percentage down-time, expected life time, failure-frequency, etc.
3. The nature of the work to be carried out: will production increase or decrease?, will product quality be affected?, will a manual worker need to be replaced by a skilled operator, and will this affect the economics of the process?
4. The extent of the automation present in the existing manufacturing system: the installation of a robot into a man-operated manufacturing line can be justified only in special circumstances.
5. The size of the company: the use of one robot may not be economically feasible but the use of a line of ten might be.

Before purchasing a particular robot system it is essential that the tasks to be performed are defined and the performance required is established. These details can then be compared with manufacturers' specifications and an appropriate robot can then be chosen.

Conclusions

A summary has been provided of the various problems that are encountered in understanding the behaviour of articulated mechanical systems used in computer-controlled robots. It has been assumed that robots are 'blind', ie incapable of acquiring information about their environment, and therefore incapable of adapting to changes. The problems associated with the interaction between a robot and its environment will be the subject of discussion in Volume 2 of this series.

Although robots can be used effectively (despite limitations in their dynamic control), there are several points which remain outstanding:

- □ What is the task classification?
- □ Is a rigid robot more suitable than a flexible robot?
- □ What is the best structure for a given task?
- □ What can a particular robot do?
- □ Must a robot be built for a specific task or be universally adaptable?

Furthered by economic necessity, research and development work is now increasing. The use of robots as tools is slowly becoming acceptable and by the end of the century their use will be vital for the survival of every manufacturer. Only then will a decline (in real terms) in manufacturing and purchasing costs be achieved.

References

1. Coiffet, P.; Rives, P. Robot recognition of the orientation of three-dimensional objects, in tasks involving automatic gripping. *RAIRO Automatic Systems, Analysis and Control* 1980, 14 (1).
2. Inagaki, S. What is the standardization for industrial robots? *The Industrial Robot* 1980, 7 (1).
3. Chavan, A; Cailleux, A. *Practical Recognition of Fossils* Masson, Paris, 1977, 2nd edn.
4. Feretti, M. The industrial robot file: first appraisals. *New Automatism* September - October, 1978.
5. Vertut, J.; Coiffet, P. Bilateral servomanipulation, MA23 in the direct mode and via optimized computer control. *Mechanism and Machine Theory* (*Journal of IFTOMM* special issue), 1976.
6. Liegeois, A.; Dombre, E. Analysis of industrial robots. Relationships between structure, performance and function. *Report of IRIA* 79102 (Project SURF), 1979.
7. Borrel, P. Manipulator behaviour models. *Applications for Performance Analysis and Automatic Control* Thesis, Montpellier, 1979.
8. Roth, B. Performance evaluation of manipulation from a kinematic point of view. *Cour de Robotique* Vol. 1, IRIA, 1976.
9. Khalil, W. *Computer Modelling and Control of the Manipulator MA23. Extension to Computer Design of Manipulators* Thesis, Montpellier, 1976.
10. Renaud, M. *Contribution to the Study of Modelling and Control of Articulated Mechanical Systems* Thesis, Toulouse, 1975.
11. Paul, R.P. The theory and practice of robot manipulators. *Programming and Control* 18th IEEE CDC Tutorial Workshop, USA, 1979.
12. Vertut, J.; Marchal, P.; Coiffet, P. The MA23 bilateral servo manipulator system. *Proceedings of 24th RSTD Conference* Boston, USA, June, 1976.
13. Bruhat, G.; Foch, A. *Mechanics* Masson, Paris, 1961, 6th edn.
14. Vertut, J.; Charles, J.; Coiffet, P.; Petit, M. Advance of the new MA23 force-reflecting manipulator system. *Robotic Manipulator Systems* 1976.
15. Oustaloup, A. Servo-control systems of the ½, 1, 3/2 and 2 range: comparative study to facilitate choice. *L'Onde Electrique* 1979, 59 (2).
16. Fournier, A. *Robotic Movement Generation. Application of Generalized Inverses and Pseudo-inverses* Thesis, Montpellier, 1980.
17. Ignatiev, M.B. *Robot Manipulator Control Algorithms* Joint Publication Research Service, Arlington, Virginia, USA, 1973.
18. Fournier, A. *Robotic Movement Generation. Application of Generalized Inverses and Pseudo-inverses* Thesis, Montpellier, 1980.
19. Liegeois, A. Control of automatic mechanical systems. *RIARO Automatic Series* 1975, 9 (2).
20. Fournier, A. *Robotic Movement Generation. Application of Generalized Inverses and Pseudo-inverses* Thesis, Montpellier, 1980.

21. Greville, T.N.E. Some applications of the pseudo-inverse of a matrix. *SIAM REVIEW, II* 1960, pp. 15-20.
22. Fournier, A. *Robotic Movement Generation. Application of Generalized Inverses and Pseudo-inverses* Thesis, Montpellier, 1980.
23. Hodder, W.W.; Margulies, G. The dynamic attitude equations for an n-body satallite. *Journal of the Astronautical Sciences* 1965, 12 (4).
24. Vukobratovic, M. *Legged Locomotion Robots and Anthropomorphic Mechanisms* Mihailo Pupin Institute, Belgrade, 1975.
25. Popov, E.P. Synthesis of robot control using dynamic models of articulated systems. *Proceedings of 6th IFAC Symposium (Control in Space)* 26-31 August, 1974.
26. Samin, J. *Dynamics of Flexible Bodies in Rotation* Thesis, Louvain-la-Neuve, 1974.
27. Ezetial, F.D.; Paynter, H. Computer representations of engineering systems involving fluid transients. *Trans. ASME* 1957, 79, 1840.
28. Karnopp, D.; Rosenberg, R.C. *System Dynamics – A Unified Approach* Wiley Interscience, 1975.
29. Vicker, J.J. Dynamic behaviour of spatial linkages. *Trans. ASME* 1968, 90.
30. Khalil, W. *Computer Modelling and Control of the Manipulator MA23. Extension to Computer Design of Manipulators* Thesis, Montpellier, 1976.
31. Nevins, T.L.; Sheridan, T.B.; Whitney, D.E.; Woodin, A.E. *The Multi-moded Remote Manipulator System* Teleoperator Arm Design Report E2720, Massachusetts Institute of Technology, USA, October, 1972.
32. Dillon, S.R. Automated equation generation and its application to problems in control. *Proceedings of JACC Conference* Austin, USA, June, 1974, pp. 572-580.
33. Khalil, W. *Contribution to Automatic Control of Manipulators, Using a Mathematical Model of the Mechanisms* Thesis, Montpellier, 1978.
34. Lebrun, M. Use of bond graphs in modelling and simulation of electro-hydraulic systems to aid design. *Carrefour sur la Robotique Industrielle* INSA, Lyon, June, 1980.
35. Liegeois, A.; Khalil, W. The dynamics of a class of electrically actuated and cable-driven manipulators. *Proceedings of IUTAM Symposium* Munich, 1977.
36. Khalil, W. *Contribution to Automatic Control of Manipulators, Using a Mathematical Model of the Mechanisms* Thesis, Montpellier, 1978.
37. Durante, C.; Prunet, F.; Dumas, J.M.; Coiffet, P. Under-optimal generalized control. *Report of ATP (Automatic Control)* CNRS, May, 1976.
38. Coiffet, P. Optimization of manipulator control structures. *Proceedings of IRIA Summer School* Toulouse, 20-29 September, 1976.
39. Dumas, J.M. *Optimal Implantation of Control Structures into Real Time* Thesis, Montpellier, 1979.
40. Ogata, A. *State Space Analysis of Control Systems* Prentice Hall, 1967.
41. Molinier, P. *Trajectory Tracking by Computer-controlled MA23 Manipulator. Demonstration of Temporal Problems in Numerical Computer Control* Thesis, Montpellier, 1977.
42. Molinier, P. *Trajectory Tracking by Computer-controlled MA23 Manipulator. Demonstration of Temporal Problems in Numerical Computer Control* Thesis, Montpellier, 1977.
43. Coiffet, P.; Liegeois, A.; Fournier, A.; Khalil, W.; Vertut, J. Computer-aided control of force-reflecting manipulators. *Proceedings of 7th IFAC Symposium (Automatic Control in Space)* May, 1976.
44. Coiffet, P.; Vertut, J.; Dombre, E. Mechanical design and computer configuration in the computer-aided manipulator control problem. *Proceedings of 12th Annual Conference on Manual Control* Urbana - Champaign, USA May, 1976.

45. Fournier, A. *Robotic Movement Generation. Application of Generalized Inverses and Pseudo-inverses* Thesis, Montpellier, 1980.
46. Kahn, M.E. *The Near-minimum Time Control of Open-loop Articulated Kinematic Chains* Thesis, Stanford University, 1970.
47. Paul, R. *Modelling Trajectory Calculation and Servoing of a Computer-controlled Arm* Advanced Research Projects Agency, Stanford University, 1973.
48. Bejczy, A.K. *Robot Arm Dynamic Control* Jet Propulsion Laboratory, NASA Technical Memorandum 33-669, February, 1974.
49. Raibert, M.H.; Horn, B.K.P. *Manipulator Control Using the Configuration Space Method* JPL Report, March, 1978.
50. Khatib, O.; Le Maitre, J.F. Dynamic control of manipulators operating in complex environments. *Proceedings of 3rd CISM/IFTOMM Symposium (The Theory and Practice of Robots and Manipulators)* Undine, Italy, September, 1978.
51. Freund, E.; Syrbe, M. Control of industrial robots by means of microprocessors. *Proceedings of IRIA Conference (New Trends in Systems Analysis)* Rocquencourt, France, 1976.
52. Freund, E. Path control for a redundant type of industrial robot. *Proceedings of 7th International Symposium on Industrial Robots* Tokyo, October, 1977.
53. Freund, E. The structure of decoupled non-linear systems. *International Journal of Control* 1975, 21 (3).
54. Zabala, J. *Control of Robot Manipulators Based on Dynamic Modelling* Thesis, Toulouse, July, 1978.
55. Khalil, W.; Leigeois, A.; Fournier, A. Dynamic control of articulated mechanical systems. *RIARO Automatic Series* 1979, 13 (2).
56. Khalil, W. *Contribution to Automatic Control of Manipulators, Using a Mathematical Model of the Mechanisms* Thesis, Montpellier, 1978.
57. Andre, P. *Cooperation Robots* Thesis, Besançon, 1980.
58. Smith, R.G.; Mitchell, T.M. A model for learning systems. *Proceedings of 5th IJCAI Symposium* Cambridge, 1977.
59. Guittet, J. The Spartacus project. *Journées de Robotique de l'IRIA* November, 1977.
60. Hill, J.W. Touch feedback and automatic control. *Proceedings of 4th International Symposium of External Control of Human Extremities* Dubrovnik, Yugoslavia, September, 1972.
61. Khalil, W. *Contribution to Automatic Control of Manipulators, Using a Mathematical Model of the Mechanisms* Thesis, Montpellier, 1978.
62. Molinier, P. *Trajectory Tracking by Computer-controlled MA23 Manipulator. Demonstration of Temporal Problems in Numerical Computer Control* Thesis, Montpellier, 1977.
63. Anon. *Programming Methods and Languages for Industrial Robots* IRIA, Rocquencourt, France, 27-29 June, 1979.
64. Brown, C.M.; Voelker, H.B. The PADL-2 project. *Production Automation Project (1) Report* University of Rochester, New York, August, 1979.
65. Latcombe, J.C. A structured analysis of programming devices for industrial robots. *Programming Methods and Languages for Industrial Robots* IRIA, Rocquencourt, France, 27-29 June, 1979.
66. Renaud, M. *Contribution to the Study of Modelling and Control of Articulated Mechanical Systems* Thesis, Toulouse, 1975.
67. Makino, H. A kinematical classification of robot manipulators. *Proceedings of 3rd Conference on Industrial Robot Technology* Nottingham, March, 1976.
68. Hamilton, W.R. *Elements of Quaternions* Chelsea Publishing Company, New York, 1969, 3rd edn.

69. Fournier, A. *Robotic Movement Generation. Application of Generalized Inverses and Pseudo-inverses* Thesis, Montpellier, 1980.
70. Johnsen, A. *Teleoperations and Human Augmentation* NASA Report SP5947, 11 March, 1969.
71. Amstadter, B.L. *Reliability Mathematics* McGraw-Hill, 1975.
72. Liegeois, A.; Dombre, E. Analysis of industrial robots. Relationships between structure, performance and function. *Report of IRIA* 79102 (Project SURF), 1979.

Index